Titu Andreescu
Oleg Mushkarov
Luchezar Stoyanov

Geometric Problems
on Maxima and Minima

Birkhäuser
Boston • Basel • Berlin

Titu Andreescu
The University of Texas at Dallas
Department of Science/
   Mathematics Education
Richardson, TX 75083
USA

Oleg Mushkarov
Bulgarian Academy of Sciences
Institute of Mathematics and Informatics
1113 Sofia
Bulgaria

Luchezar Stoyanov
The University of Western Australia
School of Mathematics and Statistics
Crawley, Perth WA 6009
Australia

Cover design by Mary Burgess.

Mathematics Subject Classification (2000): 00A07, 00A05, 00A06

Library of Congress Control Number: 2005935987

ISBN-10 0-8176-3517-3   eISBN 0-8176-4473-3
ISBN-13 978-0-8176-3517-6

Printed on acid-free paper.

©2006 Birkhäuser Boston   *Birkhäuser*

Based on the original Bulgarian edition, *Ekstremalni zadachi v geometriata*,
Narodna Prosveta, Sofia, 1989

Printed in the United States of America.   (KeS/MP)

9 8 7 6 5 4 3 2 1

www.birkhauser.com

Contents

Preface

Problems on maxima and minima arise naturally not only in science and engineering and their applications but also in daily life. A great variety of these have geometric nature: finding the shortest path between two objects satisfying certain conditions or a figure of minimal perimeter, area, or volume is a type of problem frequently met. Not surprisingly, people have been dealing with such problems for a very long time. Some of them, now regarded as famous, were dealt with by the ancient Greeks, whose intuition allowed them to discover the solutions of these problems even though for many of them they did not have the mathematical tools to provide rigorous proofs.

For example, one might mention here Heron's (first century CE) discovery that the light ray in space incoming from a point A and outgoing through a point B after reflection at a mirror α travels the shortest possible path from A to B having a common point with α.

Another famous problem, the so-called *isoperimetric problem*, was considered for example by Descartes (1596–1650): Of all plane figures with a given perimeter, find the one with greatest area. That the "perfect figure" solving the problem is the circle was known to Descartes (and possibly much earlier); however, a rigorous proof that this is indeed the solution was first given by Jacob Steiner in the nineteenth century.

A slightly different isoperimetric problem is attributed to Dido, the legendary queen of Carthage. She was allowed by the natives to purchase a piece of land on the coast of Africa "not larger than what an oxhide can surround." Cutting the oxhide into narrow strips, she made a long string with which she was supposed to surround as large as possible area on the seashore. How to do this in an optimal way is a problem closely related to the previous one, and in fact a solution is easily found once one knows the maximizing property of the circle.

Another problem that is both interesting and easy to state was posed in 1775 by I. F. Fagnano: Inscribe a triangle of minimal perimeter in a given acute-angled triangle. An elegant solution to this relatively simple "network problem" was given by Hermann Schwarz (1843–1921).

Most of these classical problems are discussed in Chapter 1, which presents several different methods for solving geometric problems on maxima and minima. One of these concerns applications of geometric transformations, e.g., reflection through a line or plane, rotation. The second is about appropriate use of inequalities. Another analytic method is the application of tools from the differential calculus. The last two methods considered in Chapter 1 are more geometric in nature; these are the method of partial variation and the tangency principle. Their names speak for themselves.

Chapter 2 is devoted to several types of geometric problems on maxima and minima that are frequently met. Here for example we discuss a variety of isoperimetric problems similar in nature to the ones mentioned above. Various distinguished points in the triangle and the tetrahedron can be described as the solutions of some specific problems on maxima or minima. Section 2.2 considers examples of this kind. An interesting type of problem, called Malfatti's problems, are contained in Section 2.3; these concern the positioning of several disks in a given figure in the plane so that the sum of the areas of the disks is maximal. Section 2.4 deals with some problems on maxima and minima arising in combinatorial geometry.

Chapter 3 collects some geometric problems on maxima and minima that could not be put into any of the first two chapters. Finally, Chapter 4 provides solutions and hints to all problems considered in the first three chapters.

Each section in the book is augmented by exercises and more solid problems for individual work. To make it easier to follow the arguments in the book a large number of figures is provided.

The present book is partly based on its Bulgarian version *Extremal Problems in Geometry*, written by O. Mushkarov and L. Stoyanov and published in 1989 (see [16]). This new version retains about half of the contents of the old one.

Altogether the book contains hundreds of geometric problems on maxima or minima. Despite the great variety of problems considered—from very old and classical ones like the ones mentioned above to problems discussed very recently in journal articles or used in various mathematics competitions around the world—the whole exposition of the book is kept at a sufficiently elementary level so that it can be understood by high-school students.

Apart from trying to be comprehensive in terms of types of problems and techniques for their solutions, we have also tried to offer various different levels of difficulty, thus making the book possible to use by people with different interests in mathematics, different abilities, and of different age groups. We hope we have achieved this to a reasonable extent.

The book reflects the experience of the authors as university teachers and as people who have been deeply involved in various mathematics competitions in different parts of the world for more than 25 years. The authors hope that the book will appeal to a wide audience of high-school students and mathematics teachers, graduate students, professional mathematicians, and puzzle enthusiasts. The book will be particularly useful to students involved in mathematics competitions around the world.

We are grateful to Svetoslav Savchev and Nevena Sabeva for helping us during the preparation of this book, and to David Kramer for the corrections and improvements he made when editing the text for publication.

Titu Andreescu
Oleg Mushkarov
Luchezar Stoyanov
September, 2005

Geometric Problems
on Maxima and Minima

Chapter 1

Methods for Finding Geometric Extrema

1.1 Employing Geometric Transformations

It is a rather common feature in solving geometric problems that the object of study undergoes some geometric transformation in order for it to be brought to a situation that is easier to deal with. In the present section this method is used to solve certain geometric problems on maxima and minima. The transformations involved are the well-known symmetry with respect to a line or a point, rotation, and dilation. Apart from this, in some space geometry problems we are going to use symmetry through a plane, rotation about a line, and space dilation. We refer the reader to [17] or [22] for general information about geometric transformations.

We begin with the well known *Heron's problem*.

Problem 1.1.1 *A line ℓ is given in the plane and two points A and B lying on the same side of ℓ. Find a point X on ℓ such that the broken line AXB has minimal length.*

Solution. Let B' be the reflection of B in ℓ (Fig. 1). By the properties of symmetry, we have $XB = XB'$ for any point X on ℓ, so

$$AX + XB = AX + XB' \geq AB'.$$

The equality occurs precisely when X is the intersection point X_0 of ℓ and the line segment AB'. Thus, for any point X on ℓ different from X_0,

$$AX + XB \geq AB' = AX_0 + X_0B,$$

which shows that X_0 is the unique solution of the problem. ♠

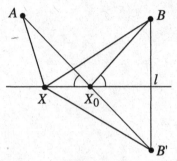

Figure 1.

The above problem shows that the shortest path from A to B having a common point with ℓ is the broken line AX_0B. It is worth mentioning that the path AX_0B satisfies the law of geometrical optics at its common point X_0 with ℓ: the angle of incidence equals the angle of reflection. It is well known from physics that this property characterizes the path of a light beam.

Problem 1.1.2 *A line ℓ is given in space and two points A and B that are not in one plane with ℓ. Find a point X on ℓ such that the broken line AXB has minimal length.*

Solution. This problem is clearly similar to Problem 1.1.1. In the solution of the latter we used symmetry with respect to a line. Notice that if α is a plane containing ℓ, the symmetry with respect to ℓ in α can be accomplished using a rotation in space through $180°$ about ℓ. Using a similar idea it is now easy to solve the present problem. Let α be the plane containing ℓ and the point A. Consider a rotation φ about ℓ that sends B to a point B' in α such that A and B' are in different half-planes of α with respect to ℓ (Fig. 2).

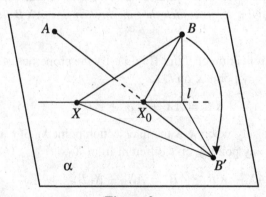

Figure 2.

If X_0 is the intersection point of ℓ and the line segment AB', for any point X on ℓ we have

$$AX + XB = AX + XB' \geq AB' = AX_0 + X_0 B',$$

with equality precisely when $X = X_0$. So the point X_0 is the unique solution of the problem.

Notice that since AX_0 and $B'X_0$ make equal angles with ℓ, the pair of line segments AX_0 and BX_0 has the same property. ♠

The main feature used in the solutions of the above two problems was that among the broken lines connecting two given points A and B the straight line segment AB has minimal length. The same elementary observation will be used in the solutions of several other problems below, while the preparation for using it will be done by means of a certain geometric transformation: symmetry, rotation, etc.

The next problem is a classic one, known as the *Schwarz triangle problem* (it is also called Fagnano's problem):

Problem 1.1.3 *Inscribe a triangle of minimal perimeter in a given acute-angled triangle.*

Solution. The next solution was given in 1900 by the Hungarian mathematician L. Fejér.

Let ABC be the given triangle. We want to find points M, N, and P on the sides BC, CA, and AB, respectively, such that the perimeter of $\triangle MNP$ is minimal.

First, we consider a simpler version of this problem. Fix an arbitrary point P on AB. We are now going to find points M and N on BC and CA, respectively, such that $\triangle MNP$ has minimal perimeter. (This minimum of course will depend on the choice of P.) Let P' be the reflection of the point P in the line BC and P'' the reflection of P in the line AC (Fig. 3 (a)). Then $CP' = CP = CP''$, $\angle P'CB = \angle PCB$, and $\angle P''CA = \angle PCA$. Setting $\gamma = \angle BCA$, we then have $\angle P'CP'' = 2\gamma$. Moreover, $2\gamma < 180°$, since $\gamma < 90°$ by assumption. Consequently, the line segment $P'P''$ intersects the sides BC and AC of $\triangle ABC$ at some points M and N, respectively, and the perimeter of $\triangle MNP$ is equal to $P'P''$. In a similar way, if X is any point on BC and Y is any point on AC, the perimeter of $\triangle XPY$ equals the length of the broken line $P'XYP''$, which is greater than or equal to $P'P''$. So, the perimeter of $\triangle PXY$ is greater than or equal to the perimeter of $\triangle PMN$, and equality holds precisely when $X = M$ and $Y = N$.

Thus, we have to find a point P on AB such that the line segment $P'P''$ has minimal length. Notice that this line segment is the base of an isosceles triangle $P''P'C$ with constant angle 2γ at C and sides $CP' = CP'' = CP$. So, we have to

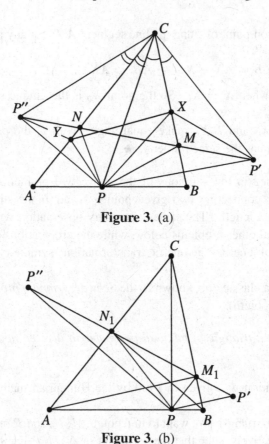

Figure 3. (a)

Figure 3. (b)

choose P on AB such that $CP' = CP$ is minimal. Obviously, for this to happen P must be the foot of the altitude through C in $\triangle ABC$.

Note now that if P is the foot of the altitude of $\triangle ABC$ through C, then M and N are the feet of the other two altitudes. To prove this, denote by M_1 and N_1 the feet of the altitudes of $\triangle ABC$ through A and B, respectively (Fig. 3 (b)). Then $\angle BM_1P' = \angle BM_1P = \angle BAC = \angle CM_1N_1$, which shows that the point P' lies on the line M_1N_1. Similarly, P'' lies on the line M_1N_1 and therefore $M = M_1$, $N = N_1$. Hence of all triangles inscribed in $\triangle ABC$, the one with vertices at the feet of the altitudes of $\triangle ABC$ has minimal perimeter. ♠

Schwarz's problem can also be solved in the case that the given triangle is not acute-angled. Assume, for example, that $\gamma \geq 90°$. It is not difficult to see that in this case the triangle MNP with minimal perimeter is such that $M = N = C$ and P is the foot of the altitude of $\triangle ABC$ through C; that is, in this case $\triangle MNP$ is degenerate.

Problem 1.1.4 *The quadrilateral in Fig. 4 is given by the coordinates of its vertices. Find the shortest path beginning at the point $A = (0, 1)$ and terminating at $C = (2, 1)$ that has common points with the sides a, d, b, d, c of the quadrilateral in this succession.*

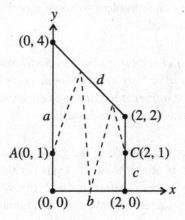

Figure 4.

Solution. Apply three successive symmetries with respect to lines as shown in Fig. 5. The image of the point C after the successive application of the three symmetries is $C' = (6, 1)$. We now want to find the shortest path from A to C' that lies entirely in the union of the quadrilaterals shown in Fig. 5. Clearly this is the broken line

$$A = (0, 1) \longrightarrow (2, 2) \longrightarrow (4, 2) \longrightarrow (6, 2) \longrightarrow C' = (6, 1).$$

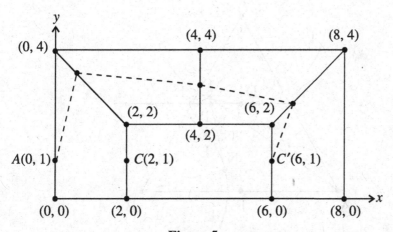

Figure 5.

Therefore the shortest path in the given quadrilateral having the desired properties
is

$$A = (0, 1) \longrightarrow (2, 2) \longrightarrow (2, 0) \longrightarrow (2, 2) \longrightarrow C = (2, 1). \spadesuit$$

We are now going to use Heron's problem to solve a problem from the 25th
International Mathematical Olympiad.

Problem 1.1.5 *A soldier has to check for mines a region having the form of an
equilateral triangle. The radius of activity of the mine detector is half the altitude
of the triangle. Assuming that the soldier starts at one of the vertices of the triangle,
find the shortest path he could use to carry out his task.*

Solution. Let h be the length of the altitude of the given equilateral $\triangle ABC$. Assume that the soldier's path starts at the point A. Consider the circles k_1 and k_2
with centers B and C, respectively, both with radius $h/2$ (Fig. 6). In order to check
the points B and C, the soldier's path must have common points with both k_1 and
k_2. Assume that the total length of the path is t and it has a common point M with
k_2 first and then a common point N with k_1. Denote by D the common point of k_2
and the altitude through C in $\triangle ABC$ and by ℓ the line through D parallel to AB.
Adding the constant $h/2$ to t and using the triangle inequality, one gets

$$t + \frac{h}{2} \geq AM + MN + NB = AM + MP + PN + NB \geq AP + PB,$$

where P is the intersection point of MN and ℓ. On the other hand, Heron's problem
(Problem 1.1.1 above) shows that $AP + PB \geq AD + DB$, where equality occurs

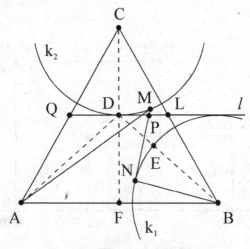

Figure 6.

precisely when $P = D$. This implies $t + \frac{h}{2} \geq AD + DB$, i.e., $t \geq AD + DE$, where E is the point of intersection of DB and k_1.

The above argument shows that the shortest path of the soldier that starts at A and has common points first with k_2 and then with k_1 is the broken line ADE. It remains to show that moving along this path, the soldier will be able to check the whole region bounded by $\triangle ABC$.

Let F, Q, and L be the midpoints of AB, AC, and BC, respectively. Since $DL < h/2$, it follows that the disk with center D and radius $h/2$ contains the whole $\triangle QLC$. In other words, from position D the soldier will be able to check the whole region bounded by $\triangle QLC$. When the soldier moves along the line segment AD he will check all points in the region bounded by the quadrilateral $AFDQ$; while moving along DE, he will check all points in the region bounded by $FBLD$.

Thus, moving along the path ADE, the soldier will be able to check the whole region bounded by $\triangle ABC$. So, ADE is one solution of the problem. Another solution is given by the path symmetric to ADE with respect to the line CD. The above arguments also show that there are no other solutions starting at A. ♠

So far, we have only used symmetry with respect to a line. In the following several problems we are going to apply some other geometric transformations.

We pass on to a problem known as *Pompeiu's theorem*.

Problem 1.1.6 *Let ABC be an equilateral triangle and P a point in its plane. Prove that there exists a triangle with sides equal to the line segments AP, BP, and CP. This triangle is degenerate if and only if P lies on the circumcircle of ABC.*

More exactly: For each point P in the plane the inequality

$$AP + BP \geq CP$$

holds true. The equality occurs if and only if P is on the arc $\overarc{AB}$ of the circumcircle of ABC.

Solution. Let, for instance, $CP \geq AP$ and $CP \geq BP$. Consider the 60° counterclockwise rotation φ about A, and let φ carry P to P'.

Then $AP = AP'$ and $\angle PAP' = 60°$, so $\triangle APP'$ is equilateral. Thus $PP' = PA$. Note also that φ carries B to C. Hence the line segment $P'C$ is the image of PB under φ; therefore $CP' = BP$. Thus $\triangle PCP'$ has sides equal to the line segments AP, BP, and CP. Because of the assumption $CP \geq AP$, $CP \geq BP$ and since $\angle APP' = 60°$, this triangle is degenerate if and only if $\angle APC = 60° =$

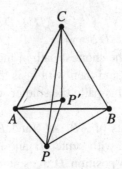

Figure 7.

$\angle ABC$, in which case $APBC$ is a cyclic quadrilateral. The latter means that the point P lies on the arc $\overset{\frown}{AB}$ of the circumcircle of ABC. ♠

The next problem is known as *Steiner's triangle problem.*

Problem 1.1.7 *Find a point X in the plane of a given triangle ABC such that the sum*

$$t(X) = AX + BX + CX$$

is minimal.

Solution. It is easy to see that if X is outside $\triangle ABC$, then there exists a point X' such that $t(X') < t(X)$. Indeed, suppose that X is exterior to the triangle. Then one of the lines AB, BC, CA, say AB, has the property that $\triangle ABC$ and the point X lie in different half-planes determined by this line (Fig. 8).

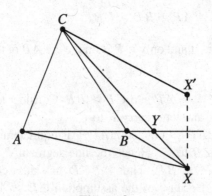

Figure 8.

Consider the reflection X' of X in AB. We have $AX' = AX$, $BX' = BX$. Also, the line segment CX intersects the line AB at some point Y, and $XY = X'Y$. Now the triangle inequality gives

$$CX' < CY + X'Y = CY + XY = CX,$$

implying $t(X') < t(X)$.

So we may restrict attention to points X in the interior or on the boundary of $\triangle ABC$. Let α, β, and γ be the angles of $\triangle ABC$. Without loss of generality we will assume that $\gamma \geq \alpha \geq \beta$. Then α and β are both acute angles.

Denote by φ the rotation through $60°$ counterclockwise about A. For any point M in the plane let $M' = \varphi(M)$. Then AMM' is an equilateral triangle. In particular, $\triangle ACC'$ is equilateral.

Consider an arbitrary point X in $\triangle ABC$. Then $AX = XX'$, while $\varphi(X) = X'$ and $\varphi(C) = C'$ imply $CX = C'X'$. Consequently, $t(X) = BX + XX' + X'C'$, i.e., $t(X)$ equals the length of the broken line $BXX'C'$.

We now consider three cases.

Case 1. $\gamma < 120°$. Then $\angle BCC' = \gamma + 60° < 180°$. Since $\alpha < 90°$, we also have $\angle BAC' < 180°$, so the line segment BC' intersects the side AC at some point D (Fig. 9 (a)). Denote by X_0 the intersection point of BC' with the circumcircle of $\triangle ACC'$. Then X_0 lies in the interior of the line segment BD and X_0' lies on $C'X_0$ since $\angle AX_0C' = \angle ACC' = 60°$.

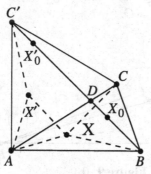

Figure 9. (a)

Moreover, we have

$$t(X_0) = BX_0 + X_0X_0' + X_0'C' = BC',$$

so $t(X_0) \leq t(X)$ for any point X in $\triangle ABC$. Equality occurs only of both X and X' lie on BC', which is possible only when $X = X_0$.

Notice that the point X_0 constructed above satisfies

$$\angle AX_0 C = \angle AX_0 B = \angle BX_0 C = 120°.$$

It is called *Torricelli's point* for $\triangle ABC$.

Case 2. $\gamma = 120°$. In this case the line segment BC' contains C and

$$t(X) = BX + XX' + X'C' = BC'$$

precisely when $X = C$.

Remark. The Cases 1 and 2 also follow by the Pompeiu theorem (Problem 1.1.6). Indeed, triangle ACC' is equilateral and we have $t(X) = AX + BX + CX \geq C'X + BX \geq C'B$.

Case 3. $\gamma > 120°$. Then BC' has no common points with the side AC (Fig. 9 (b)). If $AX \geq AC$ then the triangle inequality gives

$$t(X) = AX + BX + CX \geq AC + BC.$$

If $AX < AC$ then X' lies in $\triangle ACC'$ and

$$t(X) = BX + XX' + X'C' \geq AC + BC$$

since C lies in the rectangle $BC'X'X$ (Fig. 9 (b)). In both cases equality occurs precisely when $X = C$.

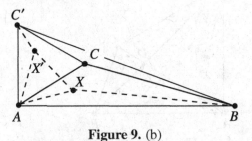

Figure 9. (b)

In conclusion, if all angles of $\triangle ABC$ are less than 120°, then $t(X)$ is minimal when X coincides with Torricelli's point of $\triangle ABC$. If one of the angles of $\triangle ABC$ is not less than 120°, then $t(X)$ is minimal when X coincides with the vertex of that angle. ◆

The following problem is a generalization of Steiner's problem.

Problem 1.1.8 *Suppose that ABC is a nonobtuse triangle, and let m, n, and p be given positive numbers. Find a point X in the plane of the triangle such that the sum*

$$s(X) = mAX + nBX + pCX$$

is minimal.

Solution. Without loss of generality we will assume that $m \geq n \geq p$.

Case 1. $m \geq n + p$. Then for any point X in the plane we have $AX + XB \geq AB$ and $AX + XC \geq AC$. Thus,

$$s(X) \geq (n + p)AX + nBX + pCX$$
$$= n(AX + XB) + p(AX + XC)$$
$$\geq nAB + pAC = s(A).$$

Moreover, it is clear that equality occurs only if $X = A$. So, the (unique) solution in this case is $X = A$.

Case 2. $m < n + p$. Then there exists a triangle $A_0B_0C_0$ with $B_0C_0 = m$, $C_0A_0 = n$, and $A_0B_0 = p$. Let α_0, β_0, and γ_0 be the angles of $\triangle A_0B_0C_0$; then $\alpha_0 \geq \beta_0 \geq \gamma_0$. Let φ be the superposition of the following two transformations: (i) the dilation with center A and ratio $k = \dfrac{p}{n}$; (ii) the rotation through angle α_0 counterclockwise about A. For any point X in the plane set $X' = \varphi(X)$ and notice that $\angle XAX' = \alpha_0 = \angle B_0A_0C_0$ (Fig. 10) and

$$\frac{AX'}{AX} = k = \frac{p}{n} = \frac{A_0B_0}{A_0C_0}.$$

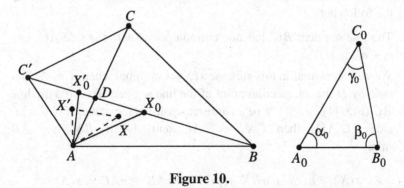

Figure 10.

Thus, $\triangle AX'X \sim \triangle A_0 B_0 C_0$, which in turn implies $\frac{XX'}{AX} = \frac{m}{n}$, i.e., $mAX = nXX'$. Also, $C'X' = kCX$, which is equivalent to $pCX = nC'X'$. Therefore, $s(X) = nXX' + nBX + nX'C'$, i.e.,

$$\frac{s(X)}{n} = XX' + BX + X'C'.$$

So, the problem is to determine X in such a way that the broken line $BXX'C'$ has minimal length.

We will now consider three subcases.

(a) The line segment BC' intersects the side AC (Fig. 10). Let D be the intersection point, and let K be the locus of the points Y in the plane such that $\angle AYD = \gamma_0$ (see Section 1.5, Example 1). Denote by X_0 the intersection point of K and the line BC. Since $\beta_0 \leq \alpha_0$, we have $\beta_0 < 90°$. This and $\beta \leq 90°$ (by assumption) gives $\beta_0 + \beta < 180°$, so B lies outside the disk determined by K. On the other hand, $\angle C'DA > \angle C'CA = \gamma_0$, so the point X_0 lies in the interior of the line segment BD. It is now clear that X'_0 lies on BC', and for any point X in the plane we have

$$\frac{s(X)}{n} \geq BC' = \frac{s(X_0)}{n},$$

where equality occurs only when $X = X_0$. Thus, in this subcase X_0 is the unique solution of the problem.

(b) The line segment BC' contains the point A. Since $A' = A$, we have $\frac{s(A)}{n} = BC'$, so $s(X)$ is minimal precisely when $X = A$.

Notice that $\gamma_0 < 90°$ and $\gamma \leq 90°$ imply $\gamma + \gamma_0 < 180°$, so BC' cannot contain the point C. So the only remaining case to consider is the following.

(c) The line segment BC' has no common points with the side AC, i.e., $\alpha + \alpha_0 > 180°$.

We will show that in this subcase $s(X)$ is minimal when $X = A$. Denote by D the intersection point of the line segment BC' and the line AC (Fig. 11), and let X be an arbitrary point in the plane. If X lies inside $\angle C'AD$, then $CX > AC$ and $AX + BX > AB$ imply

$$s(X) \geq nAX + nBX + pCX > nAB + pAC = s(A).$$

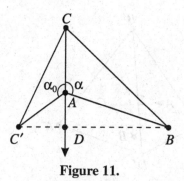

Figure 11.

If X is not in $\angle C'AD$, then the broken line $BXX'C'$ has a common point with the ray issuing from A and passing through C. Therefore

$$\frac{s(X)}{n} = BX + XX' + X'C' \geq BA + AC' = \frac{s(A)}{n},$$

where equality occurs only when $X = A$.

In conclusion, the problem always has exactly one solution. If $\alpha + \alpha_0 \geq 180°$, then $s(X)$ is minimal when $X = A$, while in the case $\alpha + \alpha_0 < 180°$, $s(X)$ is minimal when $X = X_0$. ♠

The analogues of Problems 1.1.7 and 1.1.8 for more than 3 points are no doubt very interesting. However, in general they are much more difficult. The difficulties increase substantially when one considers similar problems in space. Here we restrict ourselves to the consideration of a special case of the corresponding problem for 4 points in space.

Problem 1.1.9 *Let $ABCD$ be a regular tetrahedron in space. Find the points X in space such that the sum*

$$s(X) = AX + BX + CX + DX$$

is a minimum.

Solution. We will use the simple fact that for any point X' in a regular tetrahedron $A'B'C'D'$ the sum of the distances from X' to the four faces of the tetrahedron is constant (see below). In order to use this we construct a regular tetrahedron $A'B'C'D'$ having faces parallel to the corresponding faces of $ABCD$ and such that the point A lies in $\triangle B'C'D'$, B in $\triangle A'C'D'$, C in $\triangle A'B'D'$, and D in $\triangle A'B'C'$. The construction of such a tetrahedron is easy; just use the dilation φ with center

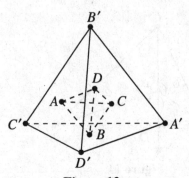

Figure 12.

O, the center of $ABCD$, and ratio $k = -3$. For any point X set $X' = \varphi(X)$. Then $A'B'C'D'$ is the desired tetrahedron (Fig. 12).

Given a point X in the tetrahedron $A'B'C'D'$, let x, y, z, and t be the distances from X to the faces of $A'B'C'D'$, and let h' be the length of its altitude. Then

$$\frac{h' \cdot [A'B'C']}{3} = \text{Vol}(A'B'C'D')$$

$$= \text{Vol}(XB'C'D') + \text{Vol}(XA'C'D')$$

$$+ \text{Vol}(XA'B'D') + \text{Vol}(XA'B'C')$$

$$= \frac{x}{3}[B'C'D'] + \frac{y}{3}[A'C'D'] + \frac{z}{3}[A'B'D'] + \frac{t}{3}[A'B'C'],$$

which gives $x + y + z + t = h'$.

If X lies outside the tetrahedron $A'B'C'D'$, then the tetrahedra $XB'C'D'$, XA' $C'D'$, $XA'B'D'$, and $XA'B'C'D'$ cover $A'B'C'D'$, so the sum of their volumes is greater than the volume of $A'B'C'D'$. So, in this case, $x + y + z + t > h'$.

To find the minimum of $s(X)$, notice that we always have $x \le XA$, where equality holds only when XA is perpendicular to the plane of triangle $B'C'D'$. Similarly, $y \le XB$, $z \le XC$, and $t \le XD$. Thus, $s(X) \ge x + y + z + t \ge h'$. Moreover, the equality $s(X) = h'$ holds if and only if X lies on the perpendiculars through A, B, C, and D to the corresponding faces of $A'B'C'D'$. Clearly the only point X with this property is $X = O$. This is the (unique) solution of the problem. ♠

The last problem in this section is quite different from the problems considered above.

Problem 1.1.10 *Given an angle Opq and a point M in its interior, draw a line through M that cuts off a triangle of minimal area from the given angle.*

Solution. It turns out that the required line ℓ is such that M is the midpoint of the line segment AB, where A and B are the intersection points of ℓ with the rays p and q, respectively. First, we construct such a line.

Let φ be the symmetry with respect to the point M. The ray $p' = \varphi(p)$ is parallel to p and intersects q at some point B_0. Let A_0 be the intersection point of p with the line MB_0. It then follows that $\varphi(A_0) = B_0$, so M is the midpoint of the line segment A_0B_0.

Next, consider an arbitrary line ℓ different from the line $\ell_0 = A_0B_0$ that intersects the rays p and q at some points A and B, respectively. We will assume that A_0 is between the points O and A; the other case is similar.

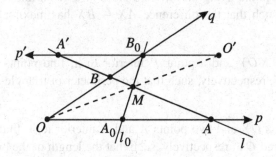

Figure 13.

Notice that $\varphi(A) = A'$, where A' is the intersection point of the ray p' and the line ℓ (Fig. 13). Thus,

$$[OAB] = [A_0MA] + [OA_0MB] = [B_0MA'] + [OA_0MB]$$
$$> [B_0BM] + [OA_0MB] = [OA_0B_0].$$

Hence the line $\ell_0 = A_0B_0$ cuts off a triangle of minimal area from the given angle. ♠

EXERCISES

1.1.11 Let M be the midpoint of the line segment AB. Show that

$$CM \le \frac{1}{2}(CA + CB),$$

for each point C. Equality occurs if and only if C lies on the line AB but outside the open line segment AB.

1.1.12 Let M and N be the midpoints of the line segments AD and BC, respectively. Show that

$$MN \leq \frac{1}{2}(AB + CD).$$

1.1.13 Find the points X lying on the boundary of a square such that the sum of distances from X to the vertices of the square is a minimum.

1.1.14 Show that of all triangles with a given base and a given area, the isosceles triangle has a minimal perimeter.

1.1.15 Let A and B be points lying on different sides of a given line ℓ. Find the points X on ℓ such that the difference $AX - BX$ has maximal absolute value.

1.1.16 Given an angle XOY and a point P interior to it, find points A and B on OX and OY, respectively, such that the perimeter of triangle PAB is a minimum.

1.1.17 Given an angle XOY and two points A and B interior to it, find points C and D on OX and OY, respectively, such that the length of the broken line $ACDB$ is a minimum.

1.1.18 Given an angle XOY and a point A on OX, find points M and N on OY and OX, respectively, such that the sum $AM + MN$ is a minimum.

1.1.19 There are given an angle with vertex A and a point P interior to it. Show how to construct a line segment BC through P with endpoints on the sides of the angle and such that

$$\frac{1}{BP} + \frac{1}{CP}$$

is a maximum.

1.1.20 Given a convex quadrilateral $ABCD$, draw a line through C, intersecting the extensions of the sides AB and AD at points M and K, such that

$$\frac{1}{[BCM]} + \frac{1}{[DCK]}$$

is a minimum.

1.1.21 An angle OXY is given and a point M in its interior. Find points A on OX and B on OY such that $OA = OB$ and the sum $MA + MB$ is a minimum.

1.1.22 Let M and N be given points in the interior of a triangle ABC. Find the shortest path starting at M and terminating at N that has common points with the sides AB, BC, and AC in this succession.

1.1.23 Let A, B, and C be three different points in the plane. Draw a line ℓ through C such that the sum of the distances from A and B to ℓ is:

(a) a minimum;   (b) a maximum.

1.1.24 Three distinct points A, B, and C are given in the plane. An arbitrary line ℓ is drawn through C, and a point M_ℓ on ℓ is chosen such that the distance sum $AM_\ell + BM_\ell$ is a minimum. What is the maximum value of the sum $AM_\ell + BM_\ell$, and for what lines ℓ is it attained?

1.1.25 Let ABC be a triangle and D, E points on the sides BC and CA such that DE passes through the incenter of ABC. Let S denote the area of the triangle CDE and r the inradius of triangle ABC. Prove that $S \geq 2r^2$.

1.1.26 In the plane of an isosceles triangle ABC with $AC = BC \geq AB$ find the points X such that the expression $r(X) = AX + BX - CX$ is a minimum.

1.1.27 Two vertices of an equilateral triangle are at distance 1 away from a point O. What is the maximum of the distance between O and the third vertex of the triangle?

1.1.28 Let ABC be a triangle with centroid G. Determine the position of the point P in the plane of ABC such that

$$AP \cdot AG + BP \cdot BG + CP \cdot CG$$

is a minimum, and express this minimum in terms of the side lengths of ABC.

1.1.29 Inscribe a quadrilateral of minimal perimeter in a given rectangle.

1.1.30 Among all quadrilaterals $ABCD$ with $AB = 3$, $CD = 2$, and $\angle AMB = 120°$, where M is the midpoint of CD, find the one of minimal perimeter.

1.1.31 Let $ABCDEF$ be a convex hexagon with $AB = BC = CD$, $DE = EF = FA$, and $\angle BCD = \angle EFA = 60°$. Let G and H be points interior to the hexagon such that the angles AGB and DHE are both $120°$. Prove that

$$AG + GB + GH + DH + HE \geq CF.$$

1.1.32 Find the points X in the plane such that the sum of the distances from X to the vertices of:

   (a) a given convex quadrilateral;

   (b) a given centrally symmetric polygon,

is a minimum.

1.1.33 Among all quadrilaterals with diagonals of given lengths and given angle between them determine the ones of minimum perimeter.

1.1.34 Let $ABCD$ be a parallelogram of area S and M a point interior to it. Prove that
$$AM \cdot CM + BM \cdot DM \geq S.$$
Determine all cases of equality if $ABCD$ is (a) a square; (b) a rectangle.

1.1.35 Let a, b, c, d be the lengths of the consecutive sides of a quadrilateral of area S. Prove that
$$S \leq \frac{1}{2}(ac + bd).$$
Equality occurs if and only if the quadrilateral is cyclic and its diagonals are perpendicular.

1.1.36 Let $ABCD$ be a tetrahedron such that $AD = BC$ and $AC = BD$. Find the points X in space such that the sum
$$t(X) = AX + BX + CX + DX$$
is a minimum.

1.1.37 Let α be a plane in space, O a given point on α, and let OA and OB be two rays on the same side of α (i.e., in the same half-space with respect to α). Find a line through O in α such that sum of the angles it makes with OA and OB is a minimum.

1.1.38 All faces of a tetrahedron $ABCD$ are acute-angled triangles. Let X, Y, Z, and T be points in the interiors of the edges AB, BC, CD, and DA, respectively. Show that:

   (a) if $\angle DAB + \angle BCD \neq \angle ABC + \angle CDA$, then among the broken lines $XYZTX$ there is none of minimum length.

   (b) if $\angle DAB + \angle BCD = \angle ABC + \angle CDA$, then there are infinitely many broken lines $XYZTX$ with a minimum length equal to $2AC \sin \frac{\alpha}{2}$, where $\alpha = \angle BAC + \angle CAD + \angle DAB$.

1.1.39 Two cities A and B are separated by a river that has parallel banks. Design a road from A to B that goes over a bridge across the river perpendicular to its banks such that the length of the road is minimal.

1.1.40 Let ABC be an equilateral triangle with side length 1. John and James play the following game. John chooses a point X on the side AC, then James chooses a point Y on BC, and finally John chooses a point Z on AB.

  (a) Suppose that John's aim is to obtain a triangle XYZ of largest possible perimeter, while James's aim is to get a triangle XYZ of smallest possible perimeter. What is the largest possible perimeter of triangle XYZ that John can achieve and with what strategy?

  (b) Suppose that John's aim is to obtain a triangle XYZ of largest possible area, while James's aim is to get a triangle XYZ of smallest possible area. What is the largest area of triangle XYZ that John can achieve and with what strategy?

1.1.41 Let $A_0 B_0 C_0$ and $A_1 B_1 C_1$ be two acute-angled triangles. Consider all triangles ABC that are similar to triangle $A_1 B_1 C_1$ (so that vertices A_1, B_1, C_1 correspond to vertices A, B, C, respectively) and circumscribed about triangle $A_0 B_0 C_0$ (where A_0 lies on BC, B_0 on CA, and C_0 on AB). Of all such possible triangles, determine the one with maximum area, and construct it.

1.2 Employing Algebraic Inequalities

A large variety of geometric problems on maxima and minima can be solved by using appropriate algebraic inequalities. Conversely, many algebraic inequalities can be interpreted geometrically as such problems. A typical example is the well-known arithmetic mean–geometric mean inequality,

$$\frac{x+y}{2} \geq \sqrt{xy} \quad (x, y \geq 0),$$

which is equivalent to the following:

    Of all rectangles with a given perimeter the square has maximal area.

In this section we solve several geometric problems on maxima and minima using classical algebraic inequalities. As one would expect, in using this approach the solution is normally given by the cases in which equality occurs. That is why it is quite important to analyze these cases carefully.

We list below some classical algebraic inequalities that are frequently used in solving geometric extremum problems.

Arithmetic Mean–Geometric Mean Inequality

For any nonnegative numbers $x_1, x_2, \ldots, x_n$,

$$\frac{x_1 + x_2 + \cdots + x_n}{n} \geq \sqrt[n]{x_1 x_2 \cdots x_n},$$

with equality if and only if $x_1 = x_2 = \cdots = x_n$.

Root Mean Square–Arithmetic Mean Inequality

For any real numbers $x_1, x_2, \ldots, x_n$,

$$\sqrt{\frac{x_1^2 + x_2^2 + \cdots + x_n^2}{n}} \geq \frac{x_1 + x_2 + \cdots + x_n}{n},$$

with equality if and only if $x_1 = x_2 = \cdots = x_n$.

Cauchy–Schwarz Inequality

For any real numbers $x_1, x_2, \ldots, x_n$ and $y_1, y_2, \ldots, y_n$,

$$(x_1^2 + x_2^2 + \cdots + x_n^2)(y_1^2 + y_2^2 + \cdots + y_n^2) \geq (x_1 y_1 + x_2 y_2 + \cdots + x_n y_n)^2,$$

with equality if and only if x_i and y_i are proportional, $i = 1, 2, \ldots, n$.

Minkowski's Inequality

For any real numbers $x_1, x_2, \ldots, x_n, y_1, y_2, \ldots, y_n, \ldots, z_1, z_2, \ldots, z_n$,

$$\sqrt{x_1^2 + y_1^2 + \cdots + z_1^2} + \sqrt{x_2^2 + y_2^2 + \cdots + z_2^2} + \cdots + \sqrt{x_n^2 + y_n^2 + \cdots + z_n^2}$$
$$\geq \sqrt{(x_1 + x_2 + \cdots + x_n)^2 + (y_1 + y_2 + \cdots + y_n)^2 + \cdots + (z_1 + z_2 + \cdots + z_n)^2},$$

with equality if and only if $x_i, y_i, \ldots, z_i$ are proportional, $i = 1, 2, \ldots, n$.

For more information on algebraic inequalities we refer the reader to the books [9], [14], [19].

We begin with the well known *isoperimetric problem for triangle*.

Problem 1.2.1 *Of all triangles with a given perimeter find the one with maximum area.*

Solution. Consider an arbitrary triangle with side lengths a, b, c and perimeter $2s = a + b + c$. By Heron's formula, its area F is given by

$$F = \sqrt{s(s - a)(s - b)(s - c)} .$$

Now the arithmetic mean–geometric mean inequality gives

$$\sqrt[3]{(s - a)(s - b)(s - c)} \le \frac{(s - a) + (s - b) + (s - c)}{3} = \frac{s}{3}.$$

Therefore

$$F \le \sqrt{s \left(\frac{s}{3} \right)^3} = s^2 \frac{\sqrt{3}}{9},$$

where equality holds if and only if $s - a = s - b = s - c$, i.e., when $a = b = c$.

Thus, the area of any triangle with perimeter $2s$ does not exceed $\frac{s^2\sqrt{3}}{9}$ and is equal to $\frac{s^2\sqrt{3}}{9}$ only for an equilateral triangle. ♠

Problem 1.2.2 *Of all rectangular boxes without a lid and having a given surface area find the one with maximum volume.*

Solution. Let x, y, and z be the edge lengths of the box (Fig. 14), and let S be its surface area.

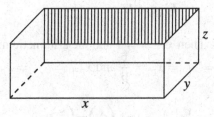

Figure 14.

Then $S = xy + 2xz + 2zy$, and the arithmetic mean–geometric mean inequality gives

$$\left(\frac{S}{3} \right)^3 = \left(\frac{xy + 2xz + 2zy}{3} \right)^3 \ge 4x^2 y^2 z^2.$$

So, for the volume $V = xyz$ of the box we get $V \le \frac{1}{2} \left(\frac{S}{3} \right)^{3/2}$. The maximum volume is obtained when equality holds, i.e., when $xy = 2xz = 2zy$. The latter easily implies that the edges of the box with maximum volume are $x = y = \sqrt{\frac{S}{3}}$ and $z = \frac{1}{2}\sqrt{\frac{S}{3}}$. ♠

The next problem is a generalization of Problem 1.1.10.

Problem 1.2.3 *Two positive integers p and q are given, and a point M in the interior of an angle with vertex O. A line through M intersects the sides of the angle at points A and B. Find the position of the line for which the product $OA^p \cdot OB^q$ is a minimum.*

Solution. Consider the points K on OA and L on OB such that MK is parallel to OB and ML is parallel to OA (Fig. 15). Then $\triangle KMA \sim \triangle OBA$ gives $OB = \frac{AB}{AM} \cdot MK$. Similarly, $OA = \frac{AB}{BM} \cdot ML$. Therefore

$$OA^p \cdot OB^q = \frac{ML^p \cdot MK^q}{\left(\frac{BM}{AB}\right)^p \cdot \left(\frac{AM}{AB}\right)^q}.$$

Since MK and ML do not depend on the choice of the line through M, it follows that $OA^p \cdot OB^q$ is minimal whenever $\left(\frac{BM}{AB}\right)^p \cdot \left(\frac{AM}{AB}\right)^q$ is maximal.

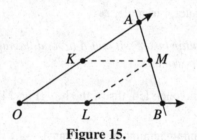

Figure 15.

Set $x = \frac{BM}{AB}$ and $y = \frac{AM}{AB}$. Then $x + y = 1$ and the arithmetic mean–geometric mean inequality for $x_1 = x_2 = \cdots = x_p = \dfrac{x}{p}$ and $x_{p+1} = \cdots = x_{p+q} = \dfrac{y}{q}$ gives

$$\frac{1}{p+q} = \frac{x+y}{p+q} \geq \sqrt[p+q]{\left(\frac{x}{p}\right)^p \left(\frac{y}{q}\right)^q}.$$

Thus $x^p \cdot y^q \leq \frac{p^p q^q}{(p+q)^{p+q}}$ and $x^p y^q$ is maximal when $\frac{x}{p} = \frac{y}{q}$, i.e., when $\frac{BM}{AM} = \frac{p}{q}$. Therefore the line through M must be drawn in such a way that $AM : MB = q : p$. Note that there exists a unique line with this property. ♠

It should be mentioned that the above problem is closely related to Problem 1.1.10 and its space analogue (see Problem 1.4.4 below). The former is obtained from Problem 1.2.3 when $p = q = 1$, while the latter uses the case $p = 1, q = 2$.

Problem 1.2.4 *Let X, Y, and Z be points on the lines determined by three pairwise skew (i.e., not lying in a plane) edges of a given cube. Find the position of these three points such that the perimeter of triangle XYZ is a minimum.*

Solution. Assume that the given cube $ABCDA_1B_1C_1D_1$ has edge of length 1. Without loss of generality we will assume that X lies on the line determined by C_1D_1, Y on the line AD and Z on the line BB_1 (Fig. 16).

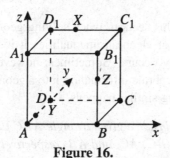

Figure 16.

Consider the coordinate system in space with origin A and coordinates axes AB, AD, and AA_1. Then the points X, Y, Z have coordinates $X = (x, 1, 1)$, $Y = (0, y, 0)$, $Z = (1, 0, z)$, and the perimeter P of $\triangle XYZ$ is given by

$$P = \sqrt{1 + y^2 + z^2} + \sqrt{(1 - x)^2 + 1 + (1 - z)^2} + \sqrt{x^2 + (1 - y)^2 + 1}.$$

Now the problem is to minimize the expression in the right-hand side when x, y, z range independently over the interval $(-\infty, +\infty)$. From its nature, one would expect this to be done by means of Minkowski's inequality. Using this inequality directly gives

$$P \geq \sqrt{[1 + (1 - x) + x]^2 + [y + (1 - y) + 1]^2 + [z + (1 - z) + 1]^2},$$

that is, $P \geq \sqrt{12}$. This may lead to the wrong conclusion that the minimum of P is $\sqrt{12}$. In fact, the above inequality is strict, i.e., equality never occurs. This can be easily derived from the condition for equality in Minkowski's inequality (see the Glossary).

Let us now show how to use Minkowski's inequality in a different way that leads to a correct result. We have

$$P \geq \sqrt{(1 + 1 + 1)^2 + (y + 1 - z + x)^2 + (z + 2 - x - y)^2}$$
$$= \sqrt{9 + (1 + x + y - z)^2 + [2 - (x + y - z)]^2}.$$

Next, using the root mean square–arithmetic mean inequality, one gets

$$(1 + x + y - z)^2 + [2 - (x + y - z)]^2 \geq \frac{9}{2},$$

and therefore $P \geq \sqrt{9 + \frac{9}{2}} = \sqrt{\frac{27}{2}}$. One checks easily that $P = \sqrt{\frac{27}{2}}$ if and only if $x = y = z = \frac{1}{2}$, showing that the perimeter of $\triangle XYZ$ is minimal precisely when X, Y, and Z are the midpoints of the corresponding edges of the cube. ♠

As we mentioned earlier in this section, when solving geometric problems on maxima and minima by means of algebraic inequalities it is rather important to investigate exactly when equality occurs. Sometimes, however, it is not an easy task to transform the obtained algebraic information into a geometric answer. Here is an example in which something similar happens.

Problem 1.2.5 *For any point X inside a given triangle ABC denote by x, y, and z the distances from X to the lines BC, AC, and AB, respectively. Find the position of X for which the sum $x^2 + y^2 + z^2$ is a minimum.*

Solution. Set $BC = a$, $CA = b$, $AB = c$. Then $2[ABC] = ax + by + cz$ and the Cauchy–Schwarz inequality gives

$$4[ABC]^2 = (ax + by + cz)^2 \leq (a^2 + b^2 + c^2)(x^2 + y^2 + z^2).$$

Hence

$$x^2 + y^2 + z^2 \geq \frac{4[ABC]^2}{a^2 + b^2 + c^2}.$$

Therefore the sum $x^2 + y^2 + z^2$ should be minimal for all points X (if any) such that

$$\frac{x}{a} = \frac{y}{b} = \frac{z}{c}.$$

What are the points X in a triangle having this property? We leave it as an exercise to the reader to find out the answer to this question. Let us just mention that for any triangle there exists only one point X satisfying the above condition. This is called *Lemoine's point*, which is defined as the intersection point of the lines symmetric to the medians of the triangle with respect to the corresponding angle bisectors.

As for the maximal value of the expression $x^2 + y^2 + z^2$, it is not difficult to see that it is achieved when X coincides with the vertex of the smallest angle of the triangle. Indeed, let $a = BC$ be the smallest side (or one of them) of $\triangle ABC$. Then $a(x + y + z) \leq ax + by + cz = 2[ABC]$, so $x + y + z \leq h_a$, where h_a is the length of the altitude through A. On the other hand, $x^2 + y^2 + z^2 \leq (x + y + z)^2$, and therefore $x^2 + y^2 + z^2 \leq h_a^2$, with equality only if $X = A$.

EXERCISES

1.2.6 Show that of all rectangles inscribed in a given circle the square has a maximum area.

1.2.7 A square is cut into several rectangles. Show that the sum of the areas of the disks determined by the circumscribed circles of these rectangles is not less than the area of the disk determined by the circumcircle of the given square.

1.2.8 Prove that of all rectangular parallelepipeds of a given volume the cube has a minimum surface area.

1.2.9 A rectangle with side lengths 1 and d is cut by two perpendicular lines into four smaller rectangles. Three of them have areas not less than 1, while the area of the fourth one is not less than 2. Find the smallest positive number d for which this is possible.

1.2.10 A square and a triangle have equal areas. Which of them has larger perimeter?

1.2.11 Find the length of the shortest line segment dividing a given triangle into two parts with equal:

(a) areas;   (b) perimeters.

1.2.12 Let O be a point in the plane of a quadrilateral $ABCD$ such that

$$AO^2 + BO^2 + CO^2 + DO^2 = 2[ABCD].$$

Prove that $ABCD$ is a square with center O.

1.2.13 A convex quadrilateral has area 1. Find the maximum of:

(a) its perimeter;   (b) the sum of its diagonals.

1.2.14 In a convex quadrilateral of area 32, the sum of the lengths of two opposite sides and one diagonal is 16. Determine all possible lengths of the other diagonal.

1.2.15 Of all tetrahedra with a right-angled trihedral angle at one of the vertices and a given sum of the six edges, find the one of maximal volume.

1.2.16 The volume and the surface area of a parallelepiped are numerically equal to 216. Prove that the parallelepiped is a cube.

1.2.17 Let α be a given plane in space and A and B two points on different sides of α. Describe the sphere through A and B that cuts off a disk of minimal area from α.

1.2.18 Let l be the length of a broken line in space, and a, b, c the lengths of its orthogonal projections onto the coordinate planes.

(a) Prove that $a + b + c \leq l\sqrt{6}$.

(b) Does there exist a closed broken line such that $a + b + c = l\sqrt{6}$?

1.2.19 For any point X in a given triangle ABC denote by x, y, and z the distances from X to the lines BC, CA, and AB, respectively. Find the position of X for which:

(a) $\dfrac{a}{x} + \dfrac{b}{y} + \dfrac{c}{z}$;   (b) $\dfrac{1}{ax} + \dfrac{1}{by} + \dfrac{1}{cz}$,

is a minimum. (Here $a = BC, b = CA, c = AB$.)

1.2.20 Let X be an arbitrary point in the interior of a tetrahedron $ABCD$ and let d_1, d_2, d_3, and d_4 be the distances from X to its faces. What is the position of X for which the product $d_1 d_2 d_3 d_4$ is a maximum?

1.2.21 Given a point X in the interior of a given triangle, one draws the lines through X parallel to the sides of the triangle. These lines divide the triangle into six parts, three of which are triangles with areas S_1, S_2, and S_3. Find the position of X such that the sum $S_1 + S_2 + S_3$ is a minimum.

1.2.22 Three lines are drawn through an interior point M of a given triangle ABC such that the first line intersects the sides AB and BC at points C_1 and A_2, the second line intersects the sides BC and CA at points A_1 and B_2, and the third line intersects the sides CA and AB at points B_1 and C_2. Find the least possible value of the sum

$$\frac{1}{[A_1 A_2 M]} + \frac{1}{[B_1 B_2 M]} + \frac{1}{[C_1 C_2 M]}.$$

1.2.23 Let X be a point in the interior of a triangle ABC and let the lines AX, BX, and CX intersect the sides BC, CA, and AB at points A_1, B_1, and C_1, respectively. Find the position of X for which the area of triangle $A_1 B_1 C_1$ is a maximum.

1.2.24 Let ABC be an equilateral triangle and P a point interior to it. Prove that the area of the triangle with sides the line segments PA, PB, and PC is not greater than $\frac{1}{3}[ABC]$.

1.2.25 Points C_1, A_1, B_1 are chosen on the sides AB, BC, CA of an equilateral triangle ABC. Determine the maximum value of the sum of the inradii of triangles $AB_1 C_1$, $BC_1 A_1$, and $CA_1 B_1$.

1.2.26 The points D and E are chosen on the sides AB and BC of a triangle ABC. The points K and M divide the line segment DE into three equal parts. The lines BK and BM intersect the side AC at T and P, respectively. Prove that $TP \le \frac{AC}{3}$.

1.2.27 Find the triangles ABC for which the expression

$$\Delta = \frac{aA + bB + cC}{a + b + c}$$

has a minimum. Does this expression have a maximum?

1.2.28 In a given sphere, inscribe a cone of maximal volume.

1.2.29 Let P be a point on a given sphere. Three mutually perpendicular rays from P meet the sphere at A, B, C. Find the maximum area of triangle ABC.

1.2.30 A trihedral angle with vertex O and a positive number a are given. Find points A, B, and C, one on its edges, such that $OA + OB + OC = a$ and the volume of the tetrahedron $OABC$ is a maximum.

1.2.31 Let M be a point lying on the base ABC of a tetrahedron $ABCD$ and let A_1, B_1, and C_1 be the feet of the perpendiculars drawn from M to the faces BCD, ACD, and ABD, respectively. Find the position of M for which the volume of the tetrahedron $MA_1B_1C_1$ is a maximum.

1.2.32 Let p, q, and r be given positive integers. A plane α passing through a given point M in the interior of a given trihedral angle with vertex O intersects its edges at points A, B, and C. Find the position of α for which the product $OA^p \cdot OB^q \cdot OC^r$ is a minimum.

1.2.33 A container having the shape of a hemisphere with radius R is full of water. A rectangualar parallelepiped with sides a and b and height $h > R$ is immersed in the container. Find the values of a and b for which such an immersion will expel a maximum volume of water from the container.

1.3 Employing Calculus

Many geometry problems on maxima and minima can be stated as problems for finding the maxima or the minima of certain functions depending on several variables. For example, the problem of inscribing a triangle of maximal area in a circle

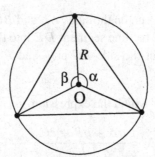

Figure 17.

is easily reduced to finding the maximum of the function $f(\alpha, \beta) = \sin \alpha \sin \beta \sin(\alpha + \beta)$, where $\alpha > 0, \beta > 0, \alpha + \beta < 180°$ (Fig. 17).

In general the function obtained in modeling the problem is complicated and difficult to investigate. Sometimes, however, one manages to reduce the problem to finding the maxima or minima of a function depending on one variable.

In this section we consider several geometric problems on maxima and minima whose solutions can be reduced to the investigation of relatively simple functions of one variable.

Before proceeding with the problems we state several facts about functions of one variable that are used in this section. Existence of extrema of functions of one variable are frequently derived by means of the well-known extreme value theorem:

Extreme Value Theorem. *If $f(t)$ is a continuous function on a finite closed interval $I = [a, b]$, then f has an (absolute) maximum and an (absolute) minimum in I.*

It is worth mentioning that f can achieve its maximal (minimal) value at more than one point. To find these points one normally uses one of the following two theorems.

Monotonicity Theorem. *Let $f(t)$ be a continuous function on an interval I and let f be differentiable in the interior of I.*

(a) *If $f(t)$ is increasing in I, then $f'(t) \geq 0$ for all t in the interior of I.*

(b) *If $f'(t) \geq 0$ for all t in the interior of I, then f is increasing in I. Moreover, if $f'(t) > 0$ for all but finitely many t in the interior of I, then f is strictly increasing in I.*

The assumption that I is an interval is essential for the validity of (b). Similarly, the inequality $f'(t) \leq 0$ characterizes decreasing functions on intervals.

As a consequence of the above theorem one gets the following.

Fermat's Theorem. *Let $f(t)$ be a differentiable function on an interval I. If f has a local maximum or minimum at some point t_0 in the interior of I, then $f'(t_0) = 0$.*

In particular, if f is continuous on an interval $[a, b]$, differentiable in (a, b), and the equation $f(t) = 0$ has no solution in (a, b), then f achieves its (absolute) maximal and minimal values at the ends of the interval and nowhere else.

Intermediate Value Theorem. *If $f(t)$ is continuous in the finite closed interval $[a, b]$ and $f(a) \cdot f(b) < 0$, then there exists at least one $t \in (a, b)$ such that $f(t) = 0$.*

We start with an example in which a quadratic function is used.

Problem 1.3.1 *Two ships travel along given directions with constant speeds. At 9 a.m. the distance between them is 20 miles, at 9:35 the distance is 15 miles, while at 9:55 the distances is 13 miles. Find the time when the distance between the ships is a minimum.*

Solution. Assume that one of the ships travels along a line g, while the second travels along a line h. First, assume that g and h intersect at some point O (Fig. 18). Denote by α the angle between the two directions of motion and by A and B the positions of the ships at 9 a.m. Set $u_1 = OA$, and $u_2 = OB$ if $\angle BOA = \alpha$, and $u_1 = -OA$ if $\angle BOA = 180° - \alpha$. Let v_1 and v_2 be the speeds of the two ships.

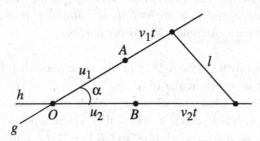

Figure 18.

Then using the law of cosines we get that the distance ℓ between the ships at time t is given by

$$\ell^2 = (u_1 + v_1 t)^2 + (u_2 + v_2 t)^2 - 2(u_1 + v_1 t)(u_2 + v_2 t) \cos \alpha.$$

Thus, ℓ^2 is a quadratic function of t, i.e., it can be written as $\ell^2 = at^2 + 2bt + c$ for some real constants a, b, and c (which can be explicitly determined by means of $u_1, u_2, v_1,$ and v_2).

In the case that g and h are parallel lines, it is also easy to see that ℓ^2 is a quadratic function of t; we leave this as an exercise to the reader.

Thus, $\ell^2 = at^2 + 2bt + c$ for some constants a, b, and c. Assume that our unit of time is 5 minutes. Then the assumptions in the problem give the following system of equations for a, b, and c:

$$400 = c,$$
$$225 = 49a + 14b + c,$$
$$169 = 121a + 22b + c.$$

The unique solution of this system is $a = 1$, $b = -16$, $c = 400$, and we have that

$$\ell^2 = t^2 - 32t + 400 = (t - 16)^2 + 144.$$

Hence $\ell \geq 12$ and $\ell = 12$ when $t = 16$. Thus the distance between the two ships is a minimum at 10:20 a.m. ♠

The following problem gives a mathematical explanation of the *law of Snell–Fermat*, well known in physics, concerning the motion of light in an inhomogeneous medium.

Problem 1.3.2 *A line ℓ is given in the plane and two points A and B on different sides of the line. A particle moves with constant speed v_1 in the half-plane containing A and with constant speed v_2 in the half-plane containing B. Find the path from A to B that is traversed in minimal time by the particle.*

Solution. Consider a coordinate system Oxy in the plane such that the axis Ox coincides with ℓ and OA is perpendicular to ℓ.

Then in coordinates, $A = (0, a)$ and $B = (d, -b)$. Without loss of generality we will assume that $a > 0$, $b > 0$, and $d > 0$ (Fig. 19). Given a point X on ℓ with coordinates $(x, 0)$, we have $AX = \sqrt{a^2 + x^2}$ and $BX = \sqrt{b^2 + (d - x)^2}$. The time t that the particle requires to traverse the broken line AXB is

$$t(x) = \frac{AX}{v_1} + \frac{BX}{v_2} = \frac{1}{v_1}\sqrt{a^2 + x^2} + \frac{1}{v_2}\sqrt{b^2 + (d - x)^2}.$$

Using a simple geometric argument, it is enough to investigate the function $t(x)$ for $0 \leq x \leq d$ (for x outside this interval $t(x)$ cannot have a minimum). We have

$$t'(x) = \frac{x}{v_1\sqrt{a^2 + x^2}} - \frac{d - x}{v_2\sqrt{b^2 + (d - x)^2}}.$$

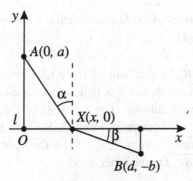

Figure 19.

It follows from

$$\frac{x}{\sqrt{a^2 + x^2}} = \frac{1}{\sqrt{\frac{a^2}{x^2} + 1}}$$

that the function $\dfrac{x}{\sqrt{a^2+x^2}}$ is strictly increasing in the interval $[0, d]$. Similarly,

$-\dfrac{d-x}{\sqrt{b^2+(d-x)^2}}$ is strictly increasing in the same interval, so $t'(x)$ is also strictly increasing in $[0, d]$. Since $t'(0) < 0$ and $t'(d) > 0$, the intermediate value theorem shows that there is a (unique) $x_0 \in (0, d)$ with $t'(x_0) = 0$. It is now clear that $t'(x) < 0$ for $x \in [0, x_0)$ and $t'(x) > 0$ for $x \in (x_0, d]$, so by the monotonicity theorem, $t(x)$ is strictly decreasing in $[0, x_0]$ and strictly increasing in $[x_0, d]$. Thus, $t(x)$ has a minimum at x_0. Notice that for the point $X_0 = (x_0, 0)$ the condition $t'(x_0) = 0$ can be written as

$$\frac{\sin \alpha}{v_1} = \frac{x_0}{v_1\sqrt{a^2 + x_0^2}} = \frac{d - x_0}{v_2\sqrt{b^2 + (d - x_0)^2}} = \frac{\sin \beta}{v_2},$$

where α is the angle between AX and Oy, while β is the angle between BX and Ox.

Hence there exists a unique point X_0 on ℓ such that the path AX_0B is traversed for a minimal time by the particle, and this point is characterized by the equation $\frac{\sin \alpha}{v_1} = \frac{\sin \beta}{v_2}$. ♠

The latter equality is called the *law of Snell–Fermat* for the diffraction of a light beam when it leaves a homogeneous medium and enters another one. This law has its fundamentals in the principle that a light beam always travels along a path that takes a minimal amount of time to traverse.

Problem 1.3.3 *Two externally tangent circles are inscribed in a given angle Opq. Find points A and D on the ray p and B and C on the ray q such that AB and*

*CD are parallel, the quadrilateral ABCD contains the two circles, and the line
segment AD has minimal length.*

Solution. Let r and R, $r < R$, be the radii of the circles and O_2 and O_1 their
centers (Fig. 20). We may assume that AB is tangent to the circle with radius R
and DC tangent to the circle with radius r. Let P and Q be the tangent points of
the two circles with AD, where P is between A and Q. Set $x = DQ$. We will now
find AD as a function of x. Since $O_1O_2 = R + r$, from the right-angled trapezoid
PO_1O_2Q one gets $PQ = 2\sqrt{Rr}$. On the other hand,

$$\angle PAO_1 = \frac{1}{2}\angle PAB = \frac{1}{2}(180° - \angle QDC) = 90° - \angle QDO_2 = \angle QO_2D,$$

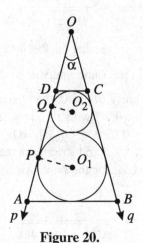

Figure 20.

so $\triangle AO_1P \sim \triangle O_2DQ$. Consequently, $\frac{R}{PA} = \frac{x}{r}$, i.e., $PA = \frac{Rr}{x}$. This implies
$AD = f(x) + 2\sqrt{Rr}$ with $f(x) = x + \frac{Rr}{x}$.

Now we have to find the minimum of $f(x)$ over the interval $0 < x < x_0 = QO$.
Notice that $\triangle PO_1O \sim \triangle QO_2O$ implies $x_0 = \frac{2r}{R-r}\sqrt{Rr}$.

We have $f'(x) = 1 - \frac{Rr}{x^2}$, and therefore $f(x)$ is strictly decreasing for $x \in$
$(0, \sqrt{Rr})$ and strictly increasing for $x \in (\sqrt{Rr}, \infty)$. Also notice that $x_0 \le \sqrt{Rr}$ is
equivalent to $3r \le R$, which in turn is equivalent to $\alpha = \angle AOB \ge 60°$.

Case 1. $3r \le R$ (i.e., $\alpha \ge 60°$). Then $x_0 \le \sqrt{Rr}$, so $f(x)$ is strictly decreasing in
$(0, x_0)$, i.e., $f(x)$ has no minimum in the interval $(0, x_0)$. In other words,
when $3r \le R$ the problem has no solution.

Case 2. $3r > R$ (i.e., $\alpha < 60°$). Then $\sqrt{Rr} < x_0$ and clearly on the interval
$(0, x_0)$, $f(x)$ has a minimum at $x = \sqrt{Rr}$. Hence the minimum length
of AD is $4\sqrt{Rr}$. The construction of the trapezoid $ABCD$ can be done
by first finding the point D on QO such that $QD = \sqrt{Rr}$. After that the
construction of the points A, B, and C is straightforward. ♠

Problem 1.3.4 *A corridor having the shape of a letter Γ is a units wide in one of
its wings and b units wide in the other. Find the length of the longest stick that can
move from one of the wings to the other. (It is assumed that the thickness of the
stick is negligible and during the motion the stick stays horizontal.)*

Solution. Consider an arbitrary angle α between $0°$ and $90°$, and let AB be a line
segment in the corner of the corridor that is tangent to the vertex O of the inside
right angle of the corridor and makes angle α with one of its walls (see Fig. 21).
Then

$$f(\alpha) = AB = AO + OB = \frac{a}{\cos \alpha} + \frac{b}{\sin \alpha}.$$

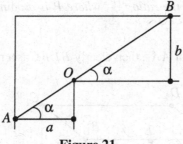

Figure 21.

A stick of length ℓ could be moved from one wing of the corridor to the other if
$\ell \leq f(\alpha)$ for all $\alpha \in (0, 90°)$.

This is a necessary and sufficient condition, so the maximal length ℓ of the stick
will be the minimum of the function $f(\alpha)$ over $(0, 90°)$ if it exists.

We have

$$f'(\alpha) = \frac{a \sin \alpha}{\cos^2 \alpha} - \frac{b \cos \alpha}{\sin^2 \alpha} = \frac{a \cos \alpha}{\sin^2 \alpha} \left(\tan^3 \alpha - \frac{b}{a} \right).$$

Since $\tan^3 \alpha$ increases strictly from 0 to ∞ when α runs from $0°$ to $90°$, there exists
a unique $\alpha_0 \in (0, 90°)$ such that $\tan^3 \alpha_0 = \frac{b}{a}$. Then $f'(\alpha_0) = 0$, and moreover,
$f'(\alpha) < 0$ for $\alpha \in (0, \alpha_0)$ and $f'(\alpha) > 0$ for $\alpha \in (\alpha_0, 90°)$. Thus, $f(x)$ has a

minimum at α_0. It follows from $\tan \alpha_0 = \sqrt[3]{\frac{b}{a}}$ that

$$\cos^2 \alpha_0 = \frac{1}{1 + \tan^2 \alpha_0} = \frac{1}{1 + \sqrt[3]{\frac{b^2}{a^2}}} = \frac{a^{2/3}}{a^{2/3} + b^{2/3}},$$

$$\sin^2 \alpha_0 = 1 - \cos^2 \alpha_0 = \frac{b^{2/3}}{a^{2/3} + b^{2/3}}.$$

So

$$f(\alpha_0) = \frac{a}{\cos \alpha_0} + \frac{b}{\sin \alpha_0} = \left(a^{2/3} + b^{2/3}\right)^{3/2}.$$

Hence the maximal length of a stick that can be moved from one wing of the corridor to the other is $\ell = \left(a^{2/3} + b^{2/3}\right)^{3/2}$. ♠

Problem 1.3.5 *The length of the edge of a cube $ABCDA_1B_1C_1D_1$ is 1. A point M is chosen on the extension of the edge AD such that D is between A and M and $AM = 2\sqrt{\frac{2}{5}}$. Let E be the midpoint of A_1B_1 and F the midpoint of DD_1. What is the maximum possible value of the ratio $\frac{MP}{PQ}$, where P is a point on AE, while Q is a point on CF?*

Solution. If L is the midpoint of AA_1, then clearly $BLFC$ is a rectangle (Fig. 22).

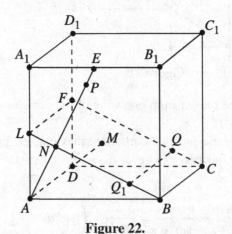

Figure 22.

For the intersection point N of AE and BL we have $\triangle ANL \sim \triangle AA_1E$, which gives $AN = \frac{1}{\sqrt{5}}$. Consider arbitrary points P and Q on the line segments AE and CF, respectively. Let Q_1 be the point on BL such that $QQ_1 \parallel BC$, and

let $x = AP - AN$ and $y = NQ_1$. We then have $PQ_1^2 = x^2 + y^2$, $PQ^2 = PQ_1^2 + QQ_1^2 = 1 + x^2 + y^2$, and $MP^2 = AM^2 + AP^2 = \frac{8}{5} + \left(\frac{1}{\sqrt{5}} + x\right)^2$.
Therefore

$$\frac{MP^2}{PQ^2} = \frac{\frac{9}{5} + \frac{2x}{\sqrt{5}} + x^2}{1 + x^2 + y^2} \le \frac{\frac{9}{5} + \frac{2x}{\sqrt{5}} + x^2}{1 + x^2},$$

where equality holds if and only if $y = 0$, that is, when $QN \parallel BC$. Clearly the latter determines the position of Q.

Since $AN = \frac{1}{\sqrt{5}}$ and $AE = \frac{\sqrt{5}}{2}$, for $x = AP - AN$ we have $x \in \Delta = \left[-\frac{1}{\sqrt{5}}, \frac{3}{2\sqrt{5}}\right]$. It remains to find the maximum of

$$f(x) = \frac{\frac{9}{5} + \frac{2x}{\sqrt{5}} + x^2}{1 + x^2} = 1 + \frac{2}{5} \cdot \frac{2 + \sqrt{5}x}{1 + x^2}$$

when $x \in \Delta$. For the function $g(x) = \frac{2+\sqrt{5}x}{1+x^2}$ we have

$$g'(x) = \frac{\sqrt{5} - 4x - \sqrt{5}x^2}{(1 + x^2)^2},$$

So $g'(1/\sqrt{5}) = 0$, $g(x)$ is strictly increasing on the interval $\left[-\frac{1}{\sqrt{5}}, \frac{1}{\sqrt{5}}\right]$, and strictly decreasing on $\left[\frac{1}{\sqrt{5}}, \frac{3}{2\sqrt{5}}\right]$. Thus $g(x)$ has a maximum $g(1/\sqrt{5}) = 5/2$ at $x = 1/\sqrt{5}$. Therefore the maximum value of $\frac{MP}{PQ}$ is $\sqrt{2}$. It is attained when $NQ \parallel BC$ and $AP = 2AN = 2/\sqrt{5}$. ◆

We have already had the opportunity to remark that not every extreme value geometric problem has a solution (see, e.g., Case 1 in Problem 1.3.4). Here we consider another problem of this kind.

Problem 1.3.6 *The length of the edge of the cube $ABCDA_1B_1C_1D_1$ is 1. Two points M and N move along the line segments AB and A_1D_1, respectively, in such a way that at any time t $(0 \le t < \infty)$ we have $BM = |\sin t|$ and $D_1N = |\sin(\sqrt{2}t)|$. Show that MN has no minimum.*

Solution. Clearly $MN \ge MA_1 \ge AA_1 = 1$. If $MN = 1$ for some t, then $M = A$ and $N = A_1$, which is equivalent to $|\sin t| = 1$ and $|\sin(\sqrt{2}t)| = 1$. Consequently $t = \frac{\pi}{2} + k\pi$ and $\sqrt{2}t = \frac{\pi}{2} + n\pi$ for some integers k and n, which implies $\sqrt{2} = \frac{2n+1}{2k+1}$, a contradiction since $\sqrt{2}$ is irrational. That is why $MN > 1$ for any t.

We will now show that MN can be made arbitrarily close to 1. For any integer k set $t_k = \frac{\pi}{2} + k\pi$. Then $|\sin t_k| = 1$, so at any time t_k the point M is at A. To show that N can be arbitrarily close to A_1 at times t_k, it is enough to show that $|\sin(\sqrt{2}t_k)|$ can be arbitrarily close to 1 for appropriate choices of k.

We are now going to use **Kronecker's theorem**: *If α is an irrational number, then the set of numbers of the form $m\alpha + n$, where m is a positive integer, while n is an arbitrary integer, is dense in the set of all real numbers.* The latter means that every nonempty open interval (regardless of how small it is) contains a number of the form $m\alpha + n$.

Since $\sqrt{2}$ is irrational, we can use Kronecker's theorem with $\alpha = \sqrt{2}$. Then for $x = \frac{1-\sqrt{2}}{2}$, and any $\delta > 0$ there exist integers $k \geq 1$ and n_k such that $k\sqrt{2} - n_k \in (x - \delta, x + \delta)$. That is, for $\epsilon_k = \sqrt{2}k + \frac{\sqrt{2}}{2} - \frac{1}{2} - n_k$ we have $|\epsilon_k| < \delta$. Since $\sqrt{2}\left(k + \frac{1}{2}\right) = \frac{1}{2} + n_k + \epsilon_k$, we have

$$\left|\sin(\sqrt{2}t_k)\right| = \left|\sin \pi \sqrt{2}\left(k + \frac{1}{2}\right)\right| = \left|\sin\left(\frac{\pi}{2} + n_k\pi + \epsilon_k\pi\right)\right| = |\cos(\pi\epsilon_k)|.$$

It remains to note that $|\cos(\delta\pi)|$ tends to 1 as δ tends to 0.

Hence MN can be made arbitrarily close to 1. ♠

EXERCISES

1.3.7 A convex quadrilateral of area S is given. Consider a parallelogram with sides parallel to the diagonals of the quadrilateral and vertices lying on its sides. Determine the maximum value of the area of such a parallelogram.

1.3.8 A point A lies between two parallel lines at distances a and b from them. Find points B and C, one on each of the lines, such that $\angle BAC = \alpha$, where $0 < \alpha < 90°$ is a given angle and the area of triangle ABC is a maximum.

1.3.9 Of all triangles inside a regular hexagon one side of which is parallel to a side of the hexagon, find those with maximal area.

1.3.10 For any triangle T denote by $S(T)$ its area and by $d(T)$ the minimal length of the diagonal of a rectangle inscribed in T. For which triangles T is the ratio $\frac{d^2(T)}{S(T)}$ a maximum?

1.3.11 A long sheet of paper having the shape of a rectangle $ABCD$ is folded along the line EF, where E is a point on the side AD and F a point on the side CD, in such a way that D is mapped to a point D' on AB (Fig. 23). What is the minimum possible area of triangle EFD?

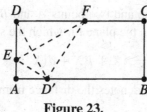

Figure 23.

1.3.12 Of all quadrilaterals inscribed in a given half-disk find the one of maximum area.

1.3.13 A convex quadrilateral of area greater than $\frac{3\sqrt{3}}{4}$ lies in a unit disk. Show that the center of the disk lies inside the quadrilateral.

1.3.14 Find a point M on the circumcircle of a right-angled triangle ABC ($\angle C = 90°$) for which the sum $MA + MB + MC$ is a maximum.

1.3.15 For any n-gon M inscribed in a unit circle k, denote by $s(M)$ the sum of the squares of its sides.

(a) Show that if $n = 3$, then the maximum value of $s(M)$ is 9 and it is attained precisely when M is an equilateral triangle.

(b) Show that if $n > 3$, then $s(M) < 9$, and for any $\epsilon > 0$ there exists an n-gon M inscribed in k with $9 - \epsilon < s(M) < 9$.

1.3.16 A regular n-gon with side a is given. One constructs a circle with center at one of the vertices of the polygon and radius less than a. Then one constructs a second circle with center at one of the neighboring vertices externally tangent to the first circle. One continues this process until circles are constructed with centers at all vertices of the polygon. Find the radius of the first circle for which the part of the polygon outside the n circles has a maximum area.

1.3.17 The vertices of an $(n + 1)$-gon lie on the sides of a regular n-gon and divide its perimeter into parts of equal length. How should one construct the $(n + 1)$-gon so that its area is:

(a) a maximum;   (b) a minimum?

1.3.18 Two points A and B lie on a given circle. Find a point C on the circle such that the sum:

(a) $AC + BC$;   (b) $AC^2 + BC^2$;   (c) $AC^3 + BC^3$

is a maximum.

1.3.19 A line ℓ is given in the plane and two points A and B on the same side of the line. Find the points X in the plane for which the sum

$$t(X) = AX + BX + d(X, \ell)$$

is a minimum. Here $d(X, \ell)$ denotes the distance from X to ℓ.

1.3.20 Four towns are the vertices of a square. Find a system of highways joining these towns such that its total length is a minimum.

1.3.21 Of all intersections of a right circular cone with planes through its vertex, find the ones of maximum area.

1.3.22 Point P lies in a given plane α, while point Q is outside α. Find a point X in α for which the ratio $d(X) = \frac{PQ + PX}{QX}$ is a maximum.

1.3.23 Given a cube $ABCDA_1B_1C_1D_1$, find the points M on the edge AB such that:

(a) the angle B_1MC_1 is a maximum;

(b) the angle A_1MC_1 is a minimum.

1.3.24 A right circular cone of volume V_1 and surface area S_1 and a circular cylinder of volume V_2 and surface area S_2 are circumscribed about the same sphere. Prove that:

(a) $3V_1 \geq 4V_2$;

(b) $4S_1 \geq (3 + 2\sqrt{2})S_2$.

1.3.25 Two balls are given in space with no common points. Find the position of a light source on the line connecting the centers of the balls such that the lighted part of the boundary spheres has a maximal total area.

1.4 The Method of Partial Variation

The method of partial variation uses the simple observation that if a function of several variables has a maximum (or a minimum) with respect to all variables, then it also has a maximum (or a minimum) with respect to any subset of variables. More precisely, assume that the function $f(x_1, x_2, \ldots, x_n)$ has a maximum (or a minimum) when $x_1 = a_1$, $x_2 = a_2$, ..., $x_n = a_n$. Then for any k, $1 \leq k < n$, the function

$$g(x_{k+1}, x_{k+2}, \ldots, x_n) = f(a_1, a_2, \ldots, a_k, x_{k+1}, x_{k+2}, \ldots, x_n)$$

has a maximum (resp. a minimum) at $x_{k+1} = a_{k+1}, x_{k+2} = a_{k+2}, \ldots, x_n = a_n$. This explanation may look a bit abstract, so let us try to explain the method of partial variation by using several examples. For more detailed discussion concerning this method and various possible applications, we refer the reader to the beautiful book of G. Polya [18].

In fact, we have already used (though implicitly) the method of partial variation in the solutions of some of the problems in the previous sections. For example, when solving Schwarz's triangle problem (see Problem 1.1.3 and Fig. 3) to inscribe a triangle MNP of minimum perimeter in a given acute triangle ABC, we did the following. We fixed a point P on AB and then found points M_P on BC and N_P on AC such that $\triangle M_P N_P P$ has a minimal perimeter. Then we found the point P on AB for which the perimeter of triangle $M_P N_P P$ is the smallest possible.

The method of partial variation can be successfully used when one knows in advance that the problem on maximum or minimum being considered has a solution. In fact, even when one does not know the existence of a solution, it is sometimes possible, using partial variation, to get some hints and even to describe precisely what the extremal object might be. For example, consider the problem to find the n-gons of maximal area among all n-gons inscribed in a given circle (this and other similar problems are considered in more detail in Section 2.1 below). Assume that there exists such an n-gon $A_1 A_2 \ldots A_n$. Fix for a moment the points $A_1, A_2, \ldots, A_{n-1}$. Then the point A_n must coincide with the midpoint A'_n of the arc $\overset{\frown}{A_{n-1} A_1}$ (Fig. 24). Indeed, if $A_n \neq A'_n$, then $[A_1 A_{n-1} A_n] < [A_1 A_{n-1} A'_n]$ and therefore the area of the polygon $A_1 A_2 \ldots A_{n-1} A_n$ is less than the area of the polygon $A_1 A_2 \ldots A_{n-1} A'_n$, a contradiction to our assumption.

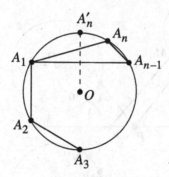

Figure 24.

Thus, using one particular partial variation we showed that $A_1 A_n = A_{n-1} A_n$. In the same way one shows that any two successive sides of the polygon must have equal lengths, so the polygon must be regular. At this point we should warn the reader that the above argument does not provide a complete solution of the problem, since we have not established the existence of an inscribed n-gon of maximal area. In many cases the existence of a solution of an extreme value geometric problem is easily derived from the extreme value theorem. However, the use of tools of this kind goes beyond the scope of this book.

In the solutions of the problems considered below we will use partial variations without assuming in advance that the respective extremal objects exist. The construction of the latter will be done in the course of the solution.

Problem 1.4.1 *A line ℓ is given in the plane and two circles k_1 and k_2 on the same side of the line. Find the shortest path from k_1 to k_2 that has a common point with ℓ.*

Solution. The problem is to find points M on k_1, N on k_2, and P on ℓ such that $t = MP + PN$ is a minimum. Fix for a moment a point M on k_1 and a point N on k_2 (Fig. 25). Then, according to Problem 1.1.1, t is a minimum when P coincides with the intersection point of ℓ and the line segment MN', where N' is the point symmetric to N through ℓ. In this case $t = MN'$. Now the problem reduces to finding the shortest line segment MN', where M is on k_1, while N' is on the symmetric image k_2' of k_2 through ℓ. If O_1 and O_2' are the centers of k_1 and k_2', then clearly the shortest such line segment is $M_0 N_0'$, where M_0 and N_0' are the intersection points of the line segment $O_1 O_2'$ with the circles k_1 and k_2', respectively.

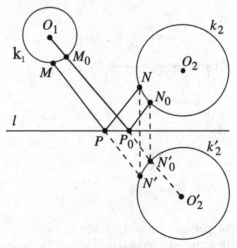

Figure 25.

Let P_0 be the intersection point of ℓ with $M_0 N_0'$ and let N_0 be the reflection of N_0' in ℓ. Then the path $M_0 P_0 N_0$ is the solution of the problem. ♠

Problem 1.4.2 *Let M be a given polygon in the plane. Show that of all triangles inscribed in M there exists one of:*

(a) *maximum area,*

(b) *maximum perimeter,*

with vertices among the vertices of M.

Solution.

(a) Consider an arbitrary $\triangle ABC$ inscribed in M, i.e., the points A, B, and C lie on the sides of M.

We have to show that there exist vertices A', B', and C' of M such that $[ABC] \leq [A'B'C']$. Fix the points A and B for a moment and let C lie on the side $C_1 C_2$ of M (Fig. 26). Then at least one of the distances from C_1 and C_2 to the line AB is not less than the distance from C to AB. Denoting the respective vertex by C' we have $[ABC'] \geq [ABC]$. In a similar way, fixing the points A and C', one finds a vertex B' of M such that $[AB'C'] \geq [ABC']$. Finally, fixing B' and C', one finds a vertex A' of M with $[A'B'C'] \geq [AB'C']$. It then follows that $[A'B'C'] \geq [ABC]$.

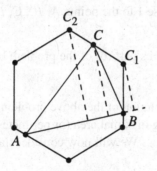

Figure 26.

Since there are only finitely many triangles with vertices among the vertices of M, there is such a triangle T of maximal area. It now follows from the above argument that any triangle inscribed in M has area not larger than the area of T.

(b) We will proceed as in (a). To do this we need the following fact.

Lemma. *Let A, B, C, and D be four different points in the plane such that C and D lie on the same side of the line AB. Then there exists a point X on CD for which the sum $AX + XB$ is a maximum and any such point coincides either with C or with D.*

Proof of the Lemma. Let ℓ be the line through C and D.

Case 1. ℓ intersects the line segment AB. Let for example D be closer to AB than C. Now we seek a point X on the line segment CD such that the broken line AXB has maximal length. Clearly the unique solution to this problem is $X = C$ (Fig. 27).

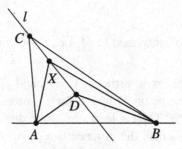

Figure 27.

Case 2. ℓ has no common points with the line segment AB. Let B' be the reflection of B in the line CD. Then $AX + XB = AX + XB'$ for any point X on CD, so we can apply Case 1 to the points A, B', C, D.

This proves the lemma.

Now using the lemma one solves part (b) of the problem by applying the same arguments as those in part (a). ♠

As we have just seen, in the solution of the above problem the task of finding a triangle inscribed in M and having maximal area (or perimeter) was reduced to the investigation of finitely many cases. We will now consider the particular case that M is a regular n-gon.

Problem 1.4.3 *Let M be a regular n-gon of area S. Find the maximum area of a triangle inscribed in M.*

Solution. According to Problem 1.4.2 above, it is enough to consider only triangles ABC with vertices among the vertices of M.

Assume that the open arcs (i.e., without their endpoints) $\overset{\frown}{AB}$, $\overset{\frown}{BC}$, and $\overset{\frown}{CA}$ contain p, q, and r vertices of M, respectively. Then $p + q + r = n - 3$. Assume

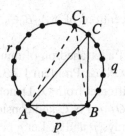

Figure 28.

that the area of triangle ABC is a maximum. We will now show that $|p - q| \leq 1$, $|p - r| \leq 1$, and $|q - r| \leq 1$. Suppose this is not the case, e.g., $q + 1 < r$. Then (see Fig. 28) if C_1 is the vertex of M next to C that is closer to A, we get $[ABC_1] > [ABC]$, a contradiction.

Therefore, setting $k = \left[\frac{n-3}{3}\right]$, we have $p = k + \epsilon_1$, $q = k + \epsilon_2$, and $r = k + \epsilon_3$, where each of the numbers ϵ_1, ϵ_2, and ϵ_3 is 0 or 1, and $\epsilon_1 + \epsilon_2 + \epsilon_3$ is the remainder of the division of n by 3. It is also clear that for such p, q, and r the area of triangle ABC is maximal. We leave it as an exercise to the reader to check that the maximum possible area of $\triangle ABC$ is

$$[ABC] = \frac{S}{n \sin(2\pi/n)} \left(\sin \frac{2(p+1)\pi}{n} + \sin \frac{2(q+1)\pi}{n} + \sin \frac{2(r+1)\pi}{n} \right).$$

This formula takes a simpler form in each of the three possible cases: $n = 3k$, $n = 3k + 1$, $n = 3k + 2$. ♠

The next problem is a space analogue of Problem 1.1.10 as well as a special case of Problem 1.2.32. The solution considered here makes use of Problem 1.2.3 (that is, the planar version of Problem 1.2.32).

Problem 1.4.4 *Given a trihedral angle and a point M in its interior, find a plane passing through M that cuts off a tetrahedron of minimum volume from the trihedral angle.*

Solution. Let $Opqr$ be the given trihedral angle. Looking at Problem 1.1.10 for analogy, one would assume that the required plane intersects the trihedral angle along a triangle with centroid at M. We will first show that a plane α_0 with this property exists. First construct a point P such that $\overrightarrow{OP} = 3\overrightarrow{OM}$. Denote by C_0 the intersection point of the ray r with the plane through P parallel to the plane Opq. Let C_0' be the intersection point of the line C_0M and the plane Opq. Clearly, $C_0M : MC_0' = 2 : 1$. Finally, construct points A_0 and B_0 on p and q, respectively, such that C_0' is the midpoint of AB (cf. the solution of Problem 1.1.10). It follows from the construction that M is the centroid of $\triangle A_0 B_0 C_0$, so the plane α_0 of this triangle has the required property.

We will now show that α_0 cuts off a tetrahedron of minimum volume from the given trihedral angle. To do this we will use Problem 1.2.3.

Let C be an arbitrary point on r different from O. Denote by C' the intersection point of the line CM with the plane Opq. Fix C and consider an arbitrary plane α through M and C that intersects the rays p and q. Let g be the line of intersection of the planes α and Opq, and let A and B be the intersection points of g with p and q, respectively. Then C' lies on g, and (having fixed C) the volume of $OABC$ will be minimal when the area of triangle OAB is a minimum. According to Problem 1.1.10, the latter occurs when the line g is such that C' is the midpoint of AB.

The above argument shows that it is enough to consider only planes α through M such that if C is the intersection point of α and r, and C' the intersection point of the line CM and the plane Opq, then CC' is a median in the triangle cut out by α from the trihedral angle.

Let r' be the ray along which the plane through r and M intersects the angle Opq (Fig. 29). Denote by φ the angle between p and q, and by ψ the angle between p and r'. If C' denotes the midpoint of AB, we have

$$OA \cdot OB = 2\frac{[OAB]}{\sin \psi} = 4\frac{[OAC']}{\sin \varphi} = \frac{2 \sin \psi}{\sin \varphi} OA \cdot OC'.$$

Thus, $OB = \frac{2 \sin \psi}{\sin \varphi} OC'$. Similarly, $OA = \frac{2 \sin(\varphi - \psi)}{\sin \varphi} OC'$, so

$$OA \cdot OB = \frac{4 \sin \psi \sin(\varphi - \psi)}{\sin^2 \varphi} OC'^2.$$

The volume of $OABC$ is proportional to the product $OA \cdot OB \cdot OC$, and the identity above shows that it is proportional to $OC \cdot OC'^2$. The problem now is to find the lines CC' through M such that C is on r, C' is on r', and $OC \cdot OC'^2$ is a minimum. It follows from Problem 1.2.3 that there exists exactly one line with this property and it is such that $CM : C'M = 2 : 1$.

Combining this with the previous arguments shows that there exists a unique plane α that cuts off a tetrahedron of minimum volume from the given trihedral angle and this plane intersects the trihedral angle along a triangle with centroid at M. ♠

We are now going to solve a classical problem using partial variation several times.

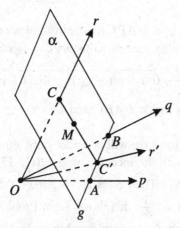

Figure 29.

Problem 1.4.5 *Show that of all tetrahedra with a given volume V the regular one has a minimum surface area.*

Solution. Fix for a moment an arbitrary $\triangle ABC$. Consider the set of points D in space such that $\text{Vol}(ABCD) = V$, i.e., the distance h from D to the plane of ABC is $h = \frac{3V}{[ABC]}$. Let D be such a point, D' its orthogonal projection on the plane of $\triangle ABC$, and x, y, and z the distances from D' to the lines BC, AC, and AB, respectively (Fig. 30).

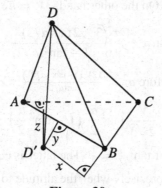

Figure 30.

For the surface area S of $ABCD$ we have

$$S = [ABC] + \frac{1}{2}\left(a\sqrt{h^2 + x^2} + b\sqrt{h^2 + y^2} + c\sqrt{h^2 + z^2}\right)$$

$$= [ABC] + \frac{1}{2}\left(\sqrt{(ah)^2 + (ax)^2} + \sqrt{(bh)^2 + (by)^2} + \sqrt{(ch)^2 + (cz)^2}\right).$$

Using the fact that $ax + by + cz \geq 2[ABC]$ (equality holds only if D' is inside $\triangle ABC$) and Minkowski's inequality (see the Glossary), one gets

$$S \geq [ABC] + \frac{1}{2}\sqrt{(ah + bh + ch)^2 + 4[ABC]^2},$$

where equality holds when D' is inside $\triangle ABC$ and $x = y = z$, i.e., if and only if D' is the incenter of $\triangle ABC$.

Using the above, in what follows we will consider only tetrahedra $ABCD$ for which the point D' coincides with the incenter of $\triangle ABC$. Fix a number $S_0 > 0$ and assume that $[ABC] = S_0$. Then $S = S_0 + \sqrt{h^2 s^2 + S_0^2}$, where s is the semiperimeter of $\triangle ABC$ and $h = \frac{3V}{S_0}$. It follows from Problem 1.2.1 that $s^2 \geq 3\sqrt{3}\, S_0$, where equality holds only for an equilateral triangle ABC. Thus

$$S \geq S_0 + \sqrt{h^2 3\sqrt{3}\, S_0 + S_0^2} = S_0 + \sqrt{27\sqrt{3}\,\frac{V^2}{S_0} + S_0^2},$$

where equality holds if and only if $\triangle ABC$ is equilateral.

The above arguments show that it is enough to consider only right triangle pyramids $ABCD$ with volume V. Given such a pyramid, denote by α the angle between a side face and the base of the pyramid. We are now going to find the value of α for which the surface area S is a minimum. Since $S = \frac{3V}{r}$, where r is the inradius of the pyramid, it is enough to find when r is a maximum. Setting $a = AB$, we have $r = \frac{a\sqrt{3}}{6} \tan \frac{\alpha}{2}$. On the other hand, $3V = hS_0 = \frac{ha^2\sqrt{3}}{4}$, and

$$h = \frac{a\sqrt{3}}{6}\tan\alpha = \frac{a\sqrt{3}}{6}\frac{2\tan(\alpha/2)}{1 - \tan^2(\alpha/2)}$$

implies $3V = \frac{a^2\tan(\alpha/2)}{4(1-\tan^2(\alpha/2))}$. Therefore $a^2 = \frac{12V\left(1-\tan^2(\alpha/2)\right)}{\tan(\alpha/2)}$. Consequently,

$$r^3 = \frac{a^3}{24\sqrt{3}}\tan^3\frac{\alpha}{2} = \frac{V}{2\sqrt{3}}\tan^2\frac{\alpha}{2}\left(1 - \tan^2\frac{\alpha}{2}\right) \leq \frac{V}{8\sqrt{3}},$$

where equality holds if and only if $\tan^2\frac{\alpha}{2} = \frac{1}{2}$. The latter is equivalent to $\cos\alpha = \frac{1-\tan^2\frac{\alpha}{2}}{1+\tan^2\frac{\alpha}{2}} = \frac{1}{3}$, which in turn holds precisely when the altitude to the base in any side face of the tetrahedron has length $\frac{a\sqrt{3}}{2}$. The latter means that all side edges of the tetrahedron have length a, i.e., that $ABCD$ is a regular tetrahedron.

Hence, for every tetrahedron with volume V and surface area S we have

$$S = \frac{3V}{r} \geq \frac{3V}{\sqrt[3]{\frac{V}{8\sqrt{3}}}} = 6\sqrt[6]{3}\, V^{2/3},$$

and equality holds if and only if it is a regular tetrahedron. ♠

EXERCISES

1.4.6 A circle k lies in the interior of a given acute angle Opq. Of all triangles MPQ, where M lies on k, P on the ray p, and Q on the ray q, find the one with minimal perimeter.

1.4.7 In a given circle k inscribe:

   (a) a triangle;   (b) a quadrilateral;   (c) a pentagon;   (d) a hexagon

of maximal area.

1.4.8 Let $ABCDEF$ be a centrally symmetric hexagon. Find points P, Q, R on its sides such that the area of triangle PQR is a maximum.

1.4.9 Let ABC be an equilateral triangle of side length 4. The points D, E, F lie on the sides BC, CA, AB, respectively, and

$$AE = BF = CD = 1.$$

The triangle QRS is formed by drawing the line segments AD, BE, and CF. For a variable point P in or on this triangle, consider the product of its distances to the three sides of ABC.

   (a) Prove that this product is a minimum when P coincides with Q, R, or S.

   (b) Determine the minimum value of this product.

1.4.10 Given three points in the plane, construct a line such that the sum of their distances to the line is a minimum.

1.4.11 Of all pentagons $ABCDE$ inscribed in a circle with radius 1 and such that $AC \perp BD$ describe the ones of minimum area.

1.4.12 Let n and p be integers such that $3 \le p < n$. Find the maximum possible area of a p-gon inscribed in a regular n-gon of area S.

1.4.13 Show that of all nondegenerate triangles ABC inscribed in a given circle there is none for which the sum $AB^3 + BC^3 + CA^3$ is a maximum. More precisely, the sum considered is a maximum if and only if two of the points A, B, C coincide and the third is their diametrically opposite point.

1.4.14 Let M be a convex polyhedron. Show that of all triangles contained in M there is one of:

   (a) maximum area;   (b) maximum perimeter

with vertices among the vertices of M.

1.4.15 Let M be a convex polyhedron. Show that of all tetrahedra contained in M and having maximal possible volume there is one whose vertices are among the vertices of M.

1.4.16 Inscribe a triangle of:

(a) maximum area;   (b) maximum perimeter

in a given cube.

1.4.17 Inscribe a tetrahedron of maximum volume in a given cube.

1.4.18 A *double quadrilateral prism* is by definition the union of two quadrilateral prisms $ABCDA_1B_1C_1D_1$ and $A_2B_2C_2D_2ABCD$ that have a common face $ABCD$ (the base of one of the prisms and the top of the other) and no other common points. Show that of all double quadrilateral prisms of a given volume the cube has a minimum surface area.

1.4.19 Show that of all tetrahedra inscribed in a given sphere the regular one has maximum volume.

1.4.20 Let $ABCD$ be a regular tetrahedron. Of all triangles LMN, where L lies on the edge AC, M in triangle ABD, and N in triangle BCD, find the one with minimal perimeter.

1.5   The Tangency Principle

This section is devoted to a method for solving geometric extremum problems using *level curves* (surfaces) of functions defined in the plane (space). The solution of the following problem gives an explanation of what this method is all about.

Problem 1.5.1 *Let ℓ be a given line in the plane and A and B two points on the same side of the line. Find a point M on ℓ such that the angle AMB is a maximum.*

Solution. It is well known that if φ is a given angle, the locus of the points M in the plane such that $\angle AMB = \varphi$ is the union of two arcs with endpoints A and B that are symmetric with respect to the line AB (Fig. 31).

Drawing these arcs for different values of φ, one gets a family of arcs covering the whole plane except the points on the line AB. Every point on the given line ℓ belongs to an arc from this family (Fig. 32), and the problem now is to find the arc having a common point with ℓ that corresponds to the largest possible value of φ.

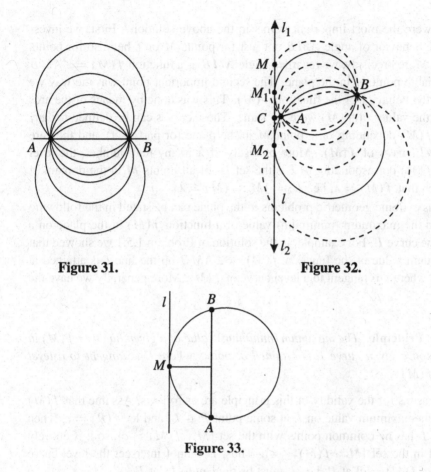

Figure 31.

Figure 32.

Figure 33.

First, consider the case that ℓ intersects the line AB. Let C be the point of intersection, and let ℓ_1 and ℓ_2 be the two rays on ℓ determined by C. Consider the arc γ_1 from the family described above that is tangent to ℓ_1, and let M_1 be the tangent point of γ_1 to ℓ_1.

Clearly γ_1 is an arc on the smallest circle through A and B that has a common point with ℓ_1. Thus, for any point M on ℓ_1 different from M_1 we have $\angle AMB < \angle AM_1B$. Similarly, the ray ℓ_2 is tangent to some arc γ_2 at some point M_2 and $\angle AMB < \angle AM_2B$ for any point M on ℓ_2 different from M_2. Now the solution of the problem is given by either M_1 or M_2 (or by both) depending on which of the angles $\angle ACM_1$ and $\angle ACM_2$ is acute. Notice that the points M_1 and M_2 are determined by the equalities $CM_1 = CM_2 = \sqrt{CA \cdot CB}$.

If $\ell \parallel AB$ then there is only one arc from the family described above that is tangent to ℓ. Hence in that case the solution is given by the intersection point M of ℓ and the perpendicular bisector of the line segment AB (Fig. 33). ♠

What were the most important points in the above solution? First, we inves-
tigated the behavior of angle AMB not just for points M on ℓ but also for points
outside ℓ. More precisely, we regarded angle AMB as a function $f(M) = \angle AMB$
of the variable point M in the plane. The second important point was the way we
looked at the behavior of the function $f(M)$. This was done by means of the arcs
on which the values of $f(M)$ were the same. These curves can be defined for any
function $f(M)$ depending on a point M in the plane (or part of it), and they are
called *level curves* of $f(M)$. More precisely, if λ is any real number, the level
curve of $f(M)$ corresponding to λ is the set L_λ of all points M in the domain of
$f(M)$ such that $f(M) = \lambda$, i.e., $L_\lambda = \{M : f(M) = \lambda\}$.

Various extreme geometric problems in the plane can be stated in the following
way: Find the maximum (minimum) value of a function $f(M)$ in the plane on a
given plane curve L. For example, in the solution of Problem 1.5.1 we showed that
the maximum value of the function $f(M) = \angle AMB$ on the line ℓ is attained at
the points where ℓ is tangent to a level curve of $f(M)$. More generally we have the
following:

Tangency Principle. *The maximum (minimum) value of a given function $f(M)$ in
the plane on a given curve L is attained at points where L is tangent to a level
curve of $f(M)$.*

The reasons for the validity of this principle are as follows. Assume that $f(M)$
achieves its maximum value on L at some point $P \in L$, and let $f(P) = c$. Then
the curve L has no common points with the set $\{M : f(M) > c\}$, so it is entirely
contained in the set $\{M : f(M) \le c\}$. Thus, L cannot intersect the level curve
$L_c = \{M : f(M) = c\}$ at P, i.e., L must be *tangent* to L_c at P.

As we saw in the solution of the Problem 1.5.1, knowing the level curves of
the function $f(M) = \angle AMB$ allowed us to easily find its extrema on the line ℓ.
Below we give various examples of functions depending on a point in the plane and
describe their level curves. For the latter in any particular problem one essentially
has to find the locus of points having a given property.

Example 1 Given two points A and B in the plane, let $f(M) = \angle AMB$. For any
$\varphi, 0 < \varphi < 180°$, the level curve L_φ of $f(M)$ is the union of two symmetric (with
respect to the line AB) arcs of circles through A and B (Fig. 34).

Example 2 Let O be a fixed point in the plane and let $f(M) = OM$. Then for
any $r > 0$ the level curve L_r is a circle with center O and radius r (Fig. 35). If we
consider points in space, then L_r is a sphere with center O and radius r.

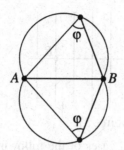

Figure 34.

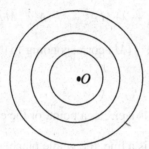

Figure 35.

Example 3 Let A and B be two fixed points in the plane and let $f(M) = MA^2 + MB^2$. Then for $r > \frac{1}{2}AB^2$, the level curve L_r is a circle with center at the midpoint O of the line segment AB (Fig. 36).

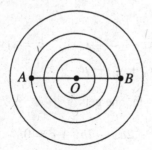

Figure 36.

Example 4 Let A and B be two fixed points in the plane and let $f(M) = MA^2 - MB^2$. Then the level curves of $f(M)$ are lines perpendicular to the line AB (Fig. 37).

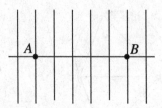

Figure 37.

The last two examples are special cases of the following more general result. Let $\lambda_1, \lambda_2, \ldots, \lambda_n$ be real numbers, and let $A_1, A_2, \ldots, A_n$ be given points in the plane. Consider the function

$$f(M) = \lambda_1 \, MA_1^2 + \lambda_2 \, MA_2^2 + \cdots + \lambda_n \, MA_n^2$$

and denote by L_μ the level curve of $f(M)$ corresponding to the real number μ.

Theorem.

(a) If $\lambda_1 + \cdots + \lambda_n \neq 0$, then L_μ is a circle, a point, or the empty set.

(b) If $\lambda_1 + \cdots + \lambda_n = 0$, then L_μ is a line, the whole plane, or the empty set.

Proof. Consider an arbitrary rectangular coordinate system Oxy in the plane, and let $M = (x, y)$ and $A_i = (x_i, y_i)$ for each $i = 1, \ldots, n$. Then $M \in L_\mu$ if and only if

$$(1) \qquad \lambda_1[(x-x_1)^2 + (y-y_1)^2] + \cdots + \lambda_n[(x-x_n)^2 + (y-y_n)^2] = \mu.$$

Set

$$\lambda = \lambda_1 + \cdots + \lambda_n, \quad a = \lambda_1 x_1 + \cdots + \lambda_n x_n, \quad b = \lambda_1 y_1 + \cdots + \lambda_n y_n,$$
$$c = \lambda_1(x_1^2 + y_1^2) + \cdots + \lambda_n(x_n^2 + y_n^2) - \mu.$$

Transforming the left-hand side of (1), one gets

$$(2) \qquad \lambda x^2 + \lambda y^2 - 2ax - 2by + c = 0.$$

If $\lambda \neq 0$, then (2) is equivalent to

$$\left(x - \frac{a}{\lambda}\right)^2 + \left(y - \frac{b}{\lambda}\right)^2 = \frac{a^2 + b^2 - \lambda c}{\lambda^2}.$$

This equation defines:

(i) a circle with center $O = (\frac{a}{\lambda}, \frac{b}{\lambda})$ if $a^2 + b^2 - \lambda c > 0$;

(ii) the point $O = (\frac{a}{\lambda}, \frac{b}{\lambda})$ if $a^2 + b^2 - \lambda c = 0$;

(iii) the empty set if $a^2 + b^2 - \lambda c < 0$.

If $\lambda = 0$, then clearly (2) defines a line if $a^2 + b^2 > 0$, the whole plane if $a = b = c = 0$, and the empty set if $a = b = 0$ and $c \neq 0$. ♠

One can prove in the same way a space analogue of the above theorem. Note that in the case (a) the level surface L_μ is a sphere, a point, or the empty set, whereas in the case (b) it is a plane, the whole space, or the empty set.

When $\lambda_1 = \lambda_2 = \cdots = \lambda_n = 1$, the above theorem gives the following:

Leibniz's Formula. *Let G be the centroid of a set of points $\{A_1, A_2, \ldots, A_n\}$ in the plane (space). Then for any point M in the plane (space) we have*

$$MA_1^2 + MA_2^2 + \cdots + MA_n^2 = n \cdots MG^2 + GA_1^2 + GA_2^2 + \cdots + GA_n^2.$$

Recall that the *centroid* of a set of points $\{A_1, A_2, \ldots, A_n\}$ is the unique point G for which $\overrightarrow{GA_1} + \overrightarrow{GA_2} + \cdots + \overrightarrow{GA_n} = \overrightarrow{0}$.

Example 5 Let G be the centroid of a set of points $\{A_1, A_2, \ldots, A_n\}$ in the plane (space) and let μ be a given number. The level curve (surface) L_μ of the function

$$f(M) = MA_1^2 + MA_2^2 + \cdots + MA_n^2$$

is a circle (sphere) with center G, the point G, or the empty set.

Example 6 Let ℓ_1 and ℓ_2 be two intersecting lines in the plane, and let $d(M, \ell_i)$ denote the distance from the point M to the line ℓ_i ($i = 1, 2$). Consider the function $f(M) = d(M, \ell_1) + d(M, \ell_2)$. The level curve L_c of $f(M)$ for $c > 0$ is the boundary of a rectangle whose diagonals lie on ℓ_1 and ℓ_2 (Fig. 38).

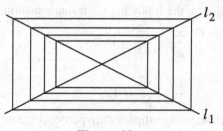

Figure 38.

The level curve L_c is easily determined using the fact that the sum of distances from an arbitrary point on the base of an isosceles triangle to the other two sides of the triangle is constant.

Example 7 Now we consider two important curves in the plane: the ellipse and the hyperbola. Let A and B be given points in the plane.

Consider the functions $f(M) = MA + MB$ and $g(M) = |MA - MB|$.

The level curves of $f(M)$ are called *ellipses*, while these of $g(M)$ are called *hyperbolas*. The points A and B are called the *foci* of these curves (Fig. 39 and Fig. 40).

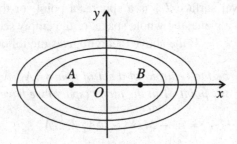

Figure 39.

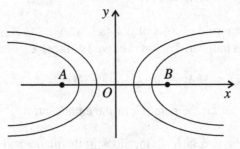

Figure 40.

Given an ellipse and a hyperbola, the line AB and its perpendicular bisector are their lines of symmetry. If one chooses these two lines for coordinates axes, then the ellipse and the hyperbola have the following Cartesian equations:

$$\frac{x^2}{a^2} + \frac{y^2}{b^2} = 1$$

and

$$\frac{x^2}{a^2} - \frac{y^2}{b^2} = 1.$$

Many interesting problems concerning ellipses or hyperbolas are related to the following main property of the tangent lines to these curves.

Focal Property. *Let M be an arbitrary point on an ellipse (hyperbola) with foci A and B. Then the segments MA and MB make equal angles with the tangent line to the ellipse (hyperbola) at the point M.*

Consider the hyperbola h given by the equation above. The lines $\ell_1 : y = \frac{b}{a} x$ and $\ell_2 : y = -\frac{b}{a} x$ are called *asymptotes* of h (Fig. 41).

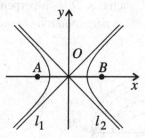

Figure 41.

It can be shown that the tangent lines to h cut off triangles of constant area from the corresponding angle between ℓ_1 and ℓ_2. This implies that the set of lines that cut off triangles of a given area from a given angle coincides with the set of tangent lines to one branch of a hyperbola with asymptotes the lines determined by the arms of the angle. Let us also mention that the tangent point of a tangent line to h coincides with the midpoint of the segment that the tangent line cuts from the angle between the asymptotes.

In what follows we consider several extreme value geometric problems and solve them using the tangency principle. The first of these problems is simple but rather instructive.

Problem 1.5.2 *Find a point M on a given line ℓ such that the distance from M to a given point O is minimal.*

Solution. We have to find the minimum value of the function $f(M) = OM$ for points M on ℓ. The level curves of $f(M)$ are circles with center O (Fig. 42).

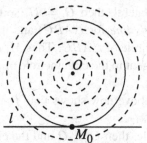

Figure 42.

Consider the level curve that is tangent to ℓ. Clearly the point of tangency M_0 gives the solution of the problem. This is actually the foot of the perpendicular from O to ℓ. ♠

Remark. One can deal in the same way with the more general problem of finding a point M on a given curve L that is closest to a given point O. In this case the solution is among the points $M_0 \in L$ such that OM_0 is perpendicular to the tangent line to L at M_0 (we then say that OM_0 is *perpendicular to the curve L*), if it exists (Fig. 43).

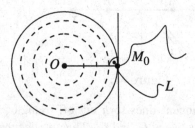

Figure 43.

More generally, if L has "corner points," one has to require that the corresponding level curve be just "touching" L at M_0. In general, the point M_0 is not unique; the reader should be able to construct examples when this happens. Another good exercise to the reader is to consider the cases in which L is a triangle, a circle, or an ellipse.

Problem 1.5.3 *Find the points M on the circumcircle of a triangle ABC such that the sum $f(M) = MA^2 + MB^2 + MC^2$ is:*
  (a) a minimum;   (b) a maximum.

Solution. It was shown in Example 5 that for any $\lambda > 0$ the level curve L_λ of $f(M)$ is a circle with center at the centroid G of $\triangle ABC$. Let O be the circumcenter of $\triangle ABC$. If $\triangle ABC$ is not equilateral, then $O \neq G$, so the line OG is well-defined (this is the so-called *Euler's line* for $\triangle ABC$) and it has two intersection points M_1 and M_2 with the circumcircle of $\triangle ABC$. Thus M_1 and M_2 are the points where a level curve of $f(M)$ is tangent to the circumcircle of $\triangle ABC$. Assume that G lies between O and M_1 (Fig. 44).

The tangency principle stated above now shows that $f(M)$ has a minimum at $M = M_1$ and a maximum at $M = M_2$.

If ABC is an equilateral triangle, then $G = O$ and the circumcircle itself is a level curve of $f(M)$, i.e., $f(M)$ is constant on it.

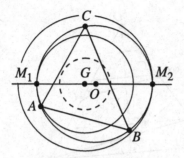

Figure 44.

In general the maximum and the minimum of $f(M)$ are easily calculated using the Leibniz formula. We leave as an exercise to the reader to show that

$$f(M_1) = \frac{1}{3}(a^2 + b^2 + c^2) + 3\left[R - \sqrt{R^2 - \frac{a^2 + b^2 + c^2}{9}}\right]^2,$$

$$f(M_2) = \frac{1}{3}(a^2 + b^2 + c^2) + 3\left[R + \sqrt{R^2 - \frac{a^2 + b^2 + c^2}{9}}\right]^2,$$

where a, b, c are the side lengths of $\triangle ABC$ and R is its circumradius. Note that the presence of a square root in these expressions yields the inequality $a^2 + b^2 + c^2 \leq 9R^2$. ♠

Problem 1.5.4 *Find the points M in the interior or on the boundary of a trapezoid $ABCD$ ($AB \parallel CD$) such that the sum of the distances from M to the sides of the trapezoid is:*
(a) a minimum,   (b) a maximum.

Solution. Denote by ℓ_1 and ℓ_2 the lines AD and BC, respectively, and by O their intersection point (Fig. 45).

Since the sum of the distances from M to AB and CD is constant, we have to find the minimum and the maximum of the function $f(M) = d(M, \ell_1) + d(M, \ell_2)$ for M running over the interior and the boundary of the trapezoid. Example 6 shows that the level curves of $f(M)$ are line segments perpendicular to the angle bisector b of angle AOB. Thus, $f(M)$ is a minimum (maximum) at the point M_1 (resp. M_2) in the trapezoid for which the distance from M_1 (resp. M_2) to the bisector b is a minimum (resp. maximum). Assume for example that $AD \leq BC$. Then clearly $M_1 = D$ and $M_2 = B$. ♠

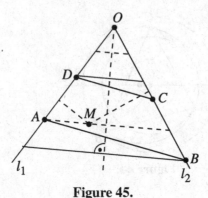

Figure 45.

In the next problem we will be seeking the extreme values of a function that depends on a variable line (instead of a point) in the plane.

Problem 1.5.5 *Let Opq be a given angle and L a given curve in its interior. Construct a line tangent to L (i.e., just touching L) that cuts off a triangle of minimal (maximal) area from the given angle.*

Solution. For any line ℓ that intersects both sides of the angle let $f(\ell)$ be the area of the triangle that ℓ cuts off from it. We have to find the minimum (maximum) of $f(\ell)$ over the set of all tangent lines ℓ to L. Following the tangency principle, we need to find the "level curves" of $f(\ell)$, i.e., the set of those lines ℓ for which $f(\ell)$ is a given constant. It is known (cf. Example 7) that the lines that cut off a triangle of a given area h from the angle Opq are tangent to one of the branches of a hyperbola with asymptotes the lines determined by p and q (Fig. 46).

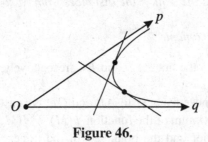

Figure 46.

Using the tangency principle, we conclude that the tangent ℓ_0 to L that cuts off a triangle of minimum (maximum) area from Opq must be tangent to L at a point M_0 at which L is tangent to a hyperbola with asymptotes the lines determined by p and q (Fig. 47). It follows from the properties of a hyperbola that M_0 is the midpoint of the line segment that the tangent line to the hyperbola at M_0 cuts from Opq. Thus the line ℓ_0 must have the same property. ♠

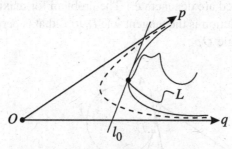

Figure 47.

Let us mention that the above argument does not guarantee the existence of a line cutting off a triangle of maximal (or minimal) area. It shows only that if such a line exists, then it must be tangent to L at a point that is the midpoint of the line segment along which the line intersects the angle. To shed a bit more light on this, let us consider two special cases.

1. Assume that L is a single point, i.e., $L = \{M\}$. Then clearly the problem about a maximum has no solution, since there are lines through M that cut off triangles of arbitrarily large area from the angle (Fig. 48).

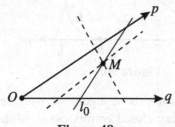

Figure 48.

Thus, in this case only the problem about the minimum makes sense. There is one line ℓ_0 through M that intersects the angle along a line segment with midpoint M, so according to the general conclusion in Problem 1.5.5, ℓ_0 cuts off a triangle of a minimum area from Opq (see Problem 1.1.10 for another proof of this fact).

2. Let k be a circle tangent to the arms p and q of the given angle at some points A and B (Fig. 49). Denote by L the smaller of the two arcs of k with endpoints A and B. Then the problem about a minimum has no solution, since the tangents to L drawn from points close to O will cut off triangles of arbitrarily small areas. (One could also say that the minimal area achievable is 0, which one gets from the lines p and q tangent to L at A and B; then the "triangles" obtained are degenerate.) The problem for maximal area has a solution, and the solution is the tangent line ℓ_0 to L that is perpendicular to the bisector of the angle Opq.

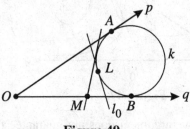

Figure 49.

Similarly, if L is the larger arc of k with endpoints A and B, then only the problem about a minimal area has a solution, and this is again the line ℓ_0 tangent to L and perpendicular to the bisector of Opq (Fig. 50).

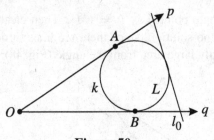

Figure 50.

Finally, let us mention that the above arguments work also in the case that k is replaced by an arbitrary closed convex curve (without corner points) inscribed in angle Opq.

EXERCISES

1.5.6 Let A and B be fixed points in the plane. Describe the level curves of the functions:

(a) $f(M) = \min\{MA, MB\}$;   (b) $f(M) = \dfrac{MA}{MB}$.

1.5.7 Among all triangles with given length ℓ of one side and given area S, determine the ones for which the product of the three altitudes is a maximum.

1.5.8 Of all triangles ABC with given lengths of the altitude through A and the median through B find the ones for which angle CAB is maximal.

1.5.9 The points A and B lie on the same side of a given line ℓ. Find a point C on ℓ such that the distance between the feet of the altitudes through A and B in triangle ABC is minimal.

1.5.10 The points A and B lie outside a given circle k. Find the points M on k such that angle AMB is:

(a) minimal;   (b) maximal.

1.5.11 Let A be a point inside a circle with center O. Find the points M on the circle such that angle OMA is maximal.

1.5.12 Find the points M on the surface of a given cube such that the angle with vertex M subtended by one of the diagonals of the cube is minimal.

1.5.13 A line l and two points A and B are given in the plane. Find the points M on l such that $AM^2 + BM^2$ is a minimum.

1.5.14 The points A and B lie on a given circle k. Find the points M on k such that:

(a) the area of triangle ABM is maximal;

(b) the sum of squares of the sides of triangle ABM is maximal;

(c) the perimeter of triangle ABM is maximal.

1.5.15 Let $A_1, A_2, \ldots, A_n$ be given points in the plane and M a set of points in the plane. Find the points X in M for which the sum $XA_1^2 + XA_2^2 + \cdots + XA_n^2$ is a minimum. Consider the cases in which M is a line segment, a line, or a circle.

1.5.16 State and solve the space version of the above problem. Consider the cases in which M is a line, a plane, or a sphere.

1.5.17 Find the points M on the incircle of a triangle ABC such that the sum $MA^2 + MB^2 + MC^2$ is:

(a) a minimum;   (b) a maximum.

1.5.18 Find the points M on the circumcircle of a triangle ABC such that the sum $MA^2 + MB^2 - 3\,MC^2$ is:

(a) a minimum;   (b) a maximum.

Consider the cases in which triangle ABC is an isosceles right triangle with $\angle ACB = 90°$ and in which triangle ABC is equilateral.

1.5.19 Let AB be a line segment parallel to a given line ℓ. Find the maximum and the minimum of the ratio $AM : BM$ as M runs over the line ℓ.

1.5.20 Let M be a set of points in the interior of an angle Opq. Find the points X in M such that the sum of distances from X to the sides of the angle is a minimum. Consider the cases in which M is a point, a line segment, a polygon, or a circle.

1.5.21 Let L be a given curve in the interior of an angle Opq. A tangent line ℓ to L intersects the ray p at a point C and the ray q at a point D. How should the line ℓ be chosen such that:

(a) $OC + OD - CD$ is a maximum;

(b) $OC + OD + CD$ is a minimum?

Consider the cases in which L is a point, a line segment, a polygon, or a circle.

1.5.22 Of all triangles with given length of a side and a given perimeter, find the one of maximum area.

1.5.23 Let G be the centroid of a triangle ABC. Determine the maximum value of the sum $\sin \angle CAG + \sin \angle CBG$.

Chapter 2

Selected Types of Geometric Extremum Problems

2.1 Isoperimetric Problems

This section is devoted to an important class of extreme value geometric problems that have attracted mathematicians' attention for a very long time. These are the so-called *isoperimetric problems*, which, as the name suggests, deal with finding the figure of maximal area among all figures of a given kind and a given perimeter. The best-known example of such a problem is the classical isoperimetric problem, where of all plane regions (bounded by a simple closed curve) with a given perimeter one wants to find the one of maximal area. Its solution is given by the so-called isoperimetric theorem, which we state in three equivalent ways.

Isoperimetric Theorem.

  (i) *Of all plane regions with a given perimeter the disk has a maximal area.*

 (ii) *Of all plane regions of a given area the disk has a minimal perimeter.*

(iii) *Let S be the area and P the perimeter of a plane region. Then $4\pi S \leq P^2$, where equality holds only when the region is a disk.*

Here is the space analogue of this theorem:

Isoperimetric Theorem in space.

  (i) *Of all solids with a given surface area the ball has a maximum volume.*

 (ii) *Of all solids with a given volume the ball has a minimum surface area.*

(iii) *Let V be the volume and S the surface area of a solid. Then $36\pi V^2 \leq S^3$,
where equality holds only when the solid is a ball.*

We are not going to discuss here the long story related to the discovery and the proof of the isoperimetric theorem. The reader can find great deal of information on this topic for example in the books [4], [6], [10], [18], and [19]. Let us just mention that even though the isoperimetric theorem has been known for a very long time, its first rigorous proof was given much later by H. A. Schwarz.

It is of course natural to ask why the isoperimetric theorem had to wait thousands of years to become a rigorous mathematical fact. Most likely one of the main reasons is that for sufficiently rigorous and clear definitions of notions like "perimeter" and "area" one needs essential use of differential and integral calculus, which was developed by Newton and Leibniz in the seventeenth century.

Our goal in this section is to prove the isoperimetric theorem for polygons in the plane.

Isoperimetric Theorem for polygons.

(i) *Of all n-gons with a given perimeter the regular n-gon has a maximum area.*

(ii) *Of all n-gons with a given area the regular n-gon has a minimum perimeter.*

(iii) *The area S and the perimeter P of any n-gon satisfy the inequality*

$$4nS\tan\frac{\pi}{n} \leq P^2,$$

where equality holds only when the n-gon is regular.

We will derive the proof of this theorem from a sequence of problems that are interesting in their own right. The first of these problems is the isoperimetric problem for circumscribed polygons.

Problem 2.1.1 *Let $n \geq 3$ be a given integer. Show that of all n-gons circumscribed about a given circle the regular n-gon has minimum area.*

Solution. The solution presented here is taken from the book [20] of L. Fejes Tóth.

Consider an arbitrary n-gon M circumscribed about a given circle k, and let $\overline{M}$ be a regular n-gon circumscribed about k (Fig. 51).

Denote by K the disk determined by the circumcircle of $\overline{M}$ and let $s_1, \ldots, s_n$ be the (equal) areas of the sectors cut off from K by the sides of $\overline{M}$. Let s_{ij} be the area of the common parts of the segments of K cut off by the ith and jth sides of

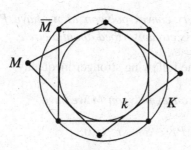

Figure 51.

M. Denote by $[M]$, $[\overline{M}]$, and $[K]$ the areas of M, $\overline{M}$, and K, respectively. Then for the area $[M \cap K]$ of the common part $M \cap K$ of M and K we have

$$[M \cap K] = [K] - (s_1 + s_2 + \cdots + s_n) + (s_{12} + s_{23} + \cdots + s_{n-1n} + s_{n1}),$$

since the total area of all parts of K lying outside M is $(s_1 + s_2 + \cdots + s_n) - (s_{12} + s_{23} + \cdots + s_{n-1n} + s_{n1})$. Therefore

$$[M] \geq [M \cap K] \geq [K] - (s_1 + s_2 + \cdots + s_n) = [\overline{M}],$$

where equality holds when no vertex of M lies in the interior of K. The latter is possible only when M is a regular n-gon, which solves the problem. ♠

Before continuing, let us introduce some notation. Let M be an arbitrary convex n-gon in the plane. Given a unit circle k_0 (i.e., a circle with radius 1) there exists a unique n-gon m circumscribed about k_0 such that the sides of m are parallel to the sides of M. This is easily seen by applying a parallel translation to each side of M until it touches the circle k_0 (Fig. 52).

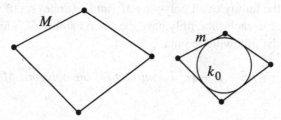

Figure 52.

Denote by S and P the *area* and *perimeter* of M, and by r the *radius* of the largest disk contained in M. The *area* of m will be denoted by s. Our next task is to prove an inequality discovered by the Swiss mathematician S. Lhuilier (1750–1840).

Lhuilier's Inequality. *For every convex polygon M we have* $P^2 \geq 4Ss$, *where equality holds if and only if M is circumscribed about a circle.*

This will be derived from the following stronger inequality.

Tóth's Inequality. *For every convex polygon M we have*

$$Pr - S - sr^2 \geq 0,$$

where equality holds if and only if M is circumscribed about a circle.

To prove the latter we will investigate the polygons M_α obtained from M by shifting its sides α units ($0 \leq \alpha \leq r$) inside the polygon keeping them parallel to their initial positions. For small α's the vertices of the polygon M_α lie on the bisectors of the corresponding angles of M (Fig. 53). Moreover, the lengths of the sides of M_α decrease when α increases, and for certain values of α some of these lengths become 0, i.e., the number of sides of the polygon M_α for such α decreases. Such values of α will be called *critical*.

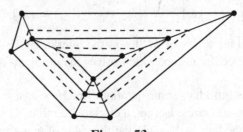

Figure 53.

The polygons M_α corresponding to critical values of α (these are given by bold lines in Fig. 53) divide the family of all polygons M_α into a (finite) set of subfamilies, so that the polygons in each subfamily have the same number of sides.

Moreover, we have the following lemma.

Lemma 1 *The expression $Pr - S - sr^2$ is constant for the polygons M_α in each subfamily.*

Proof. Suppose that M_{α_1} and M_{α_2} are polygons from the same subfamily, and let $\delta = \alpha_1 - \alpha_2 > 0$ (Fig. 54).

Then the interval (α_2, α_1) contains no critical values of α. In what follows we denote by S_i, P_i, etc., the area, perimeter, etc., of the polygon M_{α_i}. The polygon M_{α_2} comprises the following: the polygon M_{α_1}; rectangles whose bases coincide with the sides of M_{α_1} and height δ; several (as many as the number of sides of M_{α_2})

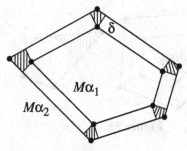

Figure 54.

additional parts, which taken together form a polygon circumscribed about a circle with radius δ and similar to m_{α_2}. Hence

$$S_2 = S_1 + P_1\delta + s_1\delta^2, \quad P_2 = P_1 + 2s_1\delta, \quad r_2 = r_1 + \delta, \quad s_2 = s_1.$$

Now a direct calculation shows that $P_2r_2 - S_2 - s_2r_2^2 = P_1r_1 - S_1 - s_1r_1^2$. This proves the lemma. ♠

The next lemma shows that when α moves across a critical value, then the expression $Pr - S - sr^2$ decreases. We continue to use the notation from the proof of Lemma 1.

Lemma 2 *Let α_0 be a critical value of α and let $\alpha_2 < \alpha_0 < \alpha_1$. Then*

$$P_2r_2 - S_2 - s_2r_2^2 > P_1r_1 - S_1 - s_1r_1^2.$$

Proof. Without loss of generality we may assume that α_0 is the only critical value of α in the interval (α_2, α_1) (explain why!). Then by Lemma 1 it is enough to prove the required inequality when $\alpha_1 = \alpha_0$ and α_2 is arbitrarily close to α_1. Let ϵ be an arbitrary positive number. Choosing α_2 sufficiently close to α_1 we have $0 < P_2 - P_1 < \epsilon$, $0 < S_2 - S_1 < \epsilon$, and $0 < r_2 - r_1 < \epsilon$. On the other hand, the fact that the number of sides of M_{α_1} is less than that of M_{α_2} (Fig. 55) implies that $s' = s_1 - s_2 > 0$. Moreover, s' does not depend on the particular choice of α_2 provided (α_2, α_1) does not contain critical values of α.
    Consequently,

$$(P_1r_1 - S_1 - s_1r_1^2) - (P_2r_2 - S_2 - s_2r_2^2)$$
$$= (P_1 - P_2)r_1 + P_2(r_1 - r_2) - (S_1 - S_2) - (s_1 - s_2)r_1^2 + s_2(r_2^2 - r_1^2).$$

Now the above inequalities yield

$$(P_1r_1 - S_1 - s_1r_1^2) - (P_2r_2 - S_2 - s_2r_2^2) < \epsilon - s'r_1^2 + \epsilon(s_1 - s')(2r_1 + \epsilon)$$

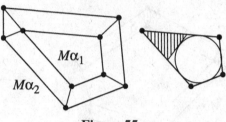

Figure 55.

for any $\epsilon > 0$. The right-hand side of the above inequality is a quadratic function of ϵ that has both a negative and a positive root. Thus we can choose $\epsilon > 0$ such that the value of this function is negative and we get $(P_1 r_1 - S_1 - s_1 r_1^2) - (P_2 r_2 - S_2 - s_2 r_2^2) < 0$, which proves the lemma. ♠

Using Lemmas 1 and 2, it is now easy to prove Tóth's inequality. Indeed, assume that $Pr - S - sr^2 < 0$. According to Lemma 1, the corresponding expression is the same for all $\alpha \in [0, \alpha')$, where α' is the first critical value of α. Then by Lemma 2, this expression gets smaller when α jumps across the critical value α', so it continues to be negative, etc. Thus, $Pr - S - sr^2 < 0$ holds for all polygons M_α for any $\alpha \in [0, r]$. However, this expression is zero when $\alpha = r$, a contradiction. Hence we always have $Pr - S - sr^2 \geq 0$.

Tóth's inequality can be written in the form $P^2 - 4sS \geq (P - 2sr)^2$, from which Lhuilier's inequality follows immediately.

It should be stressed that Lhuilier's and Tóth's inequalities are true for convex polygons only. That is why in general their application is combined with the following fact.

Problem 2.1.2 *Show that for every polygon M there exists a convex polygon M' with the same perimeter whose area is not less than the area of M.*

Solution. Let M_0 be the convex hull of M, i.e., M_0 is the smallest convex polygon containing M. Then the perimeter of M_0 is not larger than the perimeter of M. Applying a suitable dilation to M_0, one gets a polygon M' similar to M_0 and having the same perimeter as M. The area of M' is not less than the area of M since M_0 contains M. ♠

Remark. The statement in Problem 2.1.2 is true for any (bounded) region M in the plane. In many cases in dealing with isoperimetric problems this fact shows that the solution should be sought among the convex regions of the kind considered.

Proof of the Isoperimetric Theorem for n-gons. We will prove part (iii). Let M be an arbitrary n-gon. According to Problem 2.1.2, it is enough to consider the case of M convex. Then Lhuilier's inequality gives $P^2 \geq 4Ss$. Since the area of a regular n-gon circumscribed about a unit circle is equal to $n \tan \frac{\pi}{n}$, Problem 2.1.1 shows that $s \geq n \tan \frac{\pi}{n}$. Combining this with Lhuilier's inequality gives

$$P^2 \geq 4Ss \geq 4Sn \tan \frac{\pi}{n},$$

which proves the theorem. ♠

Next, we consider two interesting applications of the isoperimetric theorem and Lhuilier's inequality.

Problem 2.1.3 *Show that among all convex n-gons with given lengths of the sides the cyclic one has maximal area.*

Solution. Here we present an elegant solution of the problem given by Jacob Steiner.

We shall use without proof the fact that given an n-gon there is a unique (up to congruence) cyclic n-gon with the same sides.

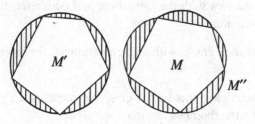

Figure 56.

Let M be an arbitrary convex polygon whose sides have the given lengths, and let M' be a cyclic polygon with the same side lengths. Let K be the disk determined by its circumcircle. On each side of M we construct externally the sector cut off from K by the respective side of M' (Fig. 56). Together with M these sectors form a region M'' whose perimeter equals the perimeter of K. Now the isoperimetric theorem implies that the area of K is not less than the area of M''. Subtracting from these two regions the sectors from K described above, we get that the area of M' is not less than that of M. ♠

Problem 2.1.4 *The area of a disk is larger than the area of any polygon with the same perimeter.*

Solution. According to Problem 2.1.2, it is enough to deal with convex polygons only. Let K and S be the areas of a disk and a (convex) polygon with the same perimeter P. Then

$$\frac{P^2}{4K} = \pi.$$

On the other hand, Lhuilier's inequality gives

$$\frac{P^2}{4S} \geq s,$$

where s is the area of a polygon circumscribed about a unit circle whose sides are parallel to the sides of M. Since the area of a unit disk is π, we have $s > \pi$, and now the above two inequalities imply $\frac{P^2}{4S} > \frac{P^2}{4K}$. Hence $K > S$, which solves the problem. ♠

EXERCISES

2.1.5 Show that if two triangles have the same base and equal perimeters, the one with a smaller (in absolute value) difference between the lengths of the other two sides has a larger area.

2.1.6 Show that of all triangles with the same base and perimeter, the isosceles triangle has maximal area.

2.1.7 Show that of all parallelograms with a given perimeter, the square has maximal area.

2.1.8 Show that of all parallelograms with a given perimeter and a given length of one of diagonals, the rhombus has maximal area.

2.1.9 Of all quadrilaterals of area 1, find the ones for which the sum of the three shortest sides is minimal.

2.1.10 Let $n \geq 3$ be an integer and $a_1, \ldots, a_{n-1}$ positive numbers. Of all n-gons $A_1 A_2 \ldots A_n$ with $A_i A_{i+1} = a_i$ for all $i = 1, \ldots, n-1$, find the ones of maximal area.

2.1.11 Let s be the length of the side of a regular n-gon inscribed in a given circle k. Show that for any nonregular n-gon M inscribed in k there exists another n-gon inscribed in k whose area is larger than that of M and that has more sides of length s than M.

2.1.12 Show that of all n-gons inscribed in a given circle the regular n-gon has a maximum area and perimeter.

2.1.13 Four congruent nonintersecting circles are centered at the vertices of a square. Construct a quadrilateral of maximum perimeter whose vertices lie on these circles.

2.1.14 Let M be a point in the interior of a convex n-gon $A_1 A_2 \ldots A_n$. Show that at least one of the angles

$$\angle M A_1 A_2, \angle M A_2 A_3, \ldots, \angle M A_{n-1} A_n, \angle M A_n A_1$$

does not exceed $\pi (n - 2)/(2n)$.

2.1.15 In a unit circle, three triangles of area 1 are drawn. Show that at least two of them have an interior point in common.

2.1.16 Show that for any nonregular n-gon there exists another n-gon with the same perimeter and of larger area whose sides have equal lengths.

2.1.17 The two ends of a rope are tied to the end of a stick. What shape should the rope take so that the device obtained in this way surrounds a region of maximal area on the ground?

2.1.18 Given a positive integer n, find a curve of a given length that cuts off a region of maximal area from an angle of measure $\frac{180°}{n}$.

2.1.19 Of all regular pyramids with n-sided bases and of a given surface area, find the ones with maximum volume.

2.1.20 Of all parallelepipeds with a given sum of the edges, find the ones with maximum volume.

2.1.21 Let a, b, c be positive numbers. Of all tetrahedra $ABCD$ with $AB = a$, $CD = b$, and $MK = c$, where M and K are the midpoints of the edges AB and CD, find the ones of maximum:

(a) surface area;   (b) volume.

2.1.22 Of all skew (i.e., nonplanar) quadrilaterals $ABCD$ in space with a given perimeter and a given side AB, find the ones for which the tetrahedron $ABCD$ has a maximum volume.

2.1.23 Of all skew quadrilaterals $ABCD$ in space with a given perimeter, find the ones for which the volume of the tetrahedron $ABCD$ is a maximum.

2.2   Extremal Points in Triangle and Tetrahedron

In every triangle (tetrahedron) there are various points defined by means of some special geometric properties. This is the way one defines the centroid, the incenter, the circumcenter, Lemoine's point, etc. It turns out that many of these points may also be characterized as the points where certain naturally defined functions in the plane or in space achieve their maxima or minima. This section is devoted to problems establishing extremal properties of the remarkable points in triangle and tetrahedron.

Note that we have already considered problems of this kind.

For example, assume that all angles of triangle ABC are less than $120°$. Recall that *Torricelli's point* of ABC is the (unique) point T with $\angle ATB = \angle BTC = \angle CTA = 120°$. According to Problem 1.1.7, the minimum of the sum $AX + BX + CX$ for points X in the plane is attained at the point T.

Here is another example. For any triangle ABC, the minimum of the sum $AX^2 + BX^2 + CX^2$ for points X in the plane (space) is attained at the centroid G of ABC. This follows from Leibniz's equality

$$AX^2 + BX^2 + CX^2 = 3XG^2 + AG^2 + BG^2 + CG^2.$$

Another extreme property of the centroid is given in Problem 1.2.23. For any point X in a triangle ABC, denote by A_1, B_1, and C_1 the intersection points of the lines AX, BX, and CX with BC, CA, and AB, respectively. Then triangle $A_1 B_1 C_1$ has maximum area when X coincides with the centroid G of ABC.

The last two properties of the centroid have analogues for a tetrahedron.

Finally, let us mention that if x, y, and z are the distances from an arbitrary point X in a given triangle ABC to its sides, then the sum $x^2 + y^2 + z^2$ is a minimum when X coincides with Lemoine's point of ABC (cf. Problem 1.2.5).

The next problem gives another extreme property of Lemoine's point.

Problem 2.2.1 *In a given triangle ABC inscribe a triangle such that the sum of the squares of its sides is minimal.*

Solution. Denote by L Lemoine's point of triangle ABC, and let M, N, and P be the orthogonal projections of L on the sides BC, AC, and AB, respectively (Fig. 57).

We are going to show that MNP (and only it) is the desired triangle.

Let us first show that L is the centroid of $\triangle MNP$. Denote by G the centroid of $\triangle MNP$ and by x_1, y_1, and z_1 the distances from G to the sides BC, AC, and AB, respectively. Then, according to Problem 1.2.5, for $x = LM$, $y = LN$, and

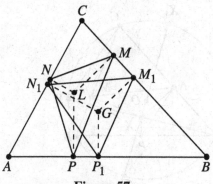

Figure 57.

$z = LP$ we have $x^2 + y^2 + z^2 \le x_1^2 + y_1^2 + z_1^2$ with equality only when $G = L$. On the other hand, Leibniz's formula for $\triangle MNP$ gives

$$x^2 + y^2 + z^2 = 3LG^2 + GM^2 + GN^2 + GP^2$$

$$\ge 3LG^2 + x_1^2 + y_1^2 + z_1^2 \ge x^2 + y^2 + z^2.$$

This gives $L = G$, i.e., L is the centroid of $\triangle MNP$.

Next, consider an arbitrary triangle $M_1 N_1 P_1$ inscribed in triangle ABC, and let G be its centroid. Denote by M_2, N_2, and P_2 the orthogonal projections of G on the sides BC, CA, and AB, respectively. Then the median formula gives

$$M_1 N_1^2 + N_1 P_1^2 + P_1 M_1^2 = 3(GM_1^2 + GN_1^2 + GP_1^2)$$

$$\ge 3(GM_2^2 + GN_2^2 + GP_2^2)$$

$$\ge 3(x^2 + y^2 + z^2) = MN^2 + NP^2 + PM^2,$$

where equality holds only when $M_1 = M_2$, $N_1 = N_2$, $P_1 = P_2$, and $G = L$, i.e., when $M_1 = M$, $N_1 = N$, and $P_1 = P$. ♠

A similar property of Lemoine's point for a tetrahedron is stated in Problem 2.2.13.

Problem 2.2.2 *Find the points X inside an acute triangle ABC such that the triangle with vertices the orthogonal projections of X on the sides of triangle ABC has maximal area.*

Solution. Let X be an arbitrary point in $\triangle ABC$, and let M, N, and P be the orthogonal projections of X on BC, CA, and AB, respectively (Fig. 58). Set $S = [ABC]$, $\sigma = [MNP]$, and let R and O be the circumradius and the circumcenter of $\triangle ABC$.

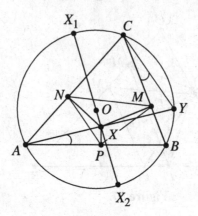

Figure 58.

We will show that σ is maximal when $X = O$. To do this we will first prove the following *Euler's formula* (cf. [8]):

$$\sigma = \left(1 - \frac{d^2}{R^2}\right) \frac{S}{4},$$

where $d = OX$.

To start the proof of this formula, write $\sigma = \frac{1}{2} MN \cdot NP \sin \angle MNP$. Denote by α, β, γ the angles of $\triangle ABC$. The quadrilateral $APXN$ is inscribed in a circle with diameter AX, so the law of sines gives $PN = AX \sin \alpha$. Similarly, from the cyclic quadrilateral $CNXM$ one finds that $MN = CX \sin \gamma$. Thus,

$$\sigma = \frac{1}{2} AX \cdot CX \sin \alpha \, \sin \gamma \, \sin \angle MNP.$$

Let Y be the intersection point of the ray AX with the circumcircle k of $\triangle ABC$. We claim that $\angle MNP = \angle XCY$. Indeed, from the quadrilateral $APXN$ one gets $\angle XNP = \angle XAP$. On the other hand, $\angle XAP = \angle YAB = \angle YCB$. It now follows from the quadrilateral $CNXM$ that $\angle XNM = \angle XCM$, so

$$\angle MNP = \angle MNX + \angle XNP = \angle XCM + \angle YCB = \angle XCY.$$

Next, notice that $\angle XYC = \angle AYC = \angle ABC = \beta$. Combining this with the above and with the law of sines for $\triangle XYC$, we get

$$\frac{CX}{XY} = \frac{\sin \beta}{\sin \angle XCY} = \frac{\sin \beta}{\sin \angle MNP}.$$

Thus $CX \sin \angle MNP = XY \sin \beta$ and one obtains

$$\sigma = \frac{1}{2} AX \cdot XY \sin \alpha \, \sin \beta \, \sin \gamma.$$

Let $X_1 X_2$ be the diameter in k containing the point X. Assume that O is between X_1 and X. Then $X_1 X_2 = R + d$ and $X_2 X = R - d$. Since the chords AY and $X_1 X_2$ intersect at X, we have

$$AX \cdot XY = X_1 X \cdot X X_2 = (R + d)(R - d) = R^2 - d^2$$

and the identity above gives

$$\sigma = \frac{R^2 - d^2}{2} \sin \alpha \, \sin \beta \, \sin \gamma.$$

On the other hand,

$$S = \frac{ab}{2} \sin \gamma = \frac{1}{2}(2R \sin \alpha)(2R \sin \beta) \sin \gamma = 2R^2 \sin \alpha \, \sin \beta \, \sin \gamma.$$

Hence

$$\sigma = \frac{R^2 - d^2}{2} \cdot \frac{S}{2R^2} = \left(1 - \frac{d^2}{R^2}\right) \frac{S}{4},$$

and Euler's formula is proved.

It is clear that σ is maximal when $d = 0$, i.e., when $X = O$. ♠

Let us note that using an argument similar to the one in the solution of Problem 2.2.2, one can show that for any point X in the plane we have

$$\sigma = \pm \left(1 - \frac{d^2}{R^2}\right) \frac{S}{4},$$

where the sign $+$ corresponds to the case that X is inside the circumcircle of $\triangle ABC$ and the sign $-$ to the case that X is outside the circumcircle. When X is on the circumcircle we have $\sigma = 0$, i.e., the points M, N, and P lie on a line. The latter fact is known as *Simson's theorem*. Further, Euler's formula shows that for any $\sigma_0 > 0$ the locus of the points X in the plane for which $\sigma = \sigma_0$ is

(i) a circle with center O and radius $R\sqrt{1 + \frac{4\sigma_0}{S}}$ if $4\sigma_0 > S$;

(ii) the union of two concentric circles with center O and radii $R\sqrt{1 + \frac{4\sigma_0}{S}}$ and $R\sqrt{1 - \frac{4\sigma_0}{S}}$ if $4\sigma_0 \leq S$.

The next two problems are taken from the article [11] of G. Lawden.

Problem 2.2.3 *Let ABC be a given triangle, and let A' be a point in the plane different from A, B, and C. Let L and M be the feet of the perpendiculars drawn from A to the lines A'B and A'C, respectively. Find the position of A' such that the length of LM is maximal.*

Solution. We will show that the length of LM is maximal when A' coincides with the center of the excircle (external circle) for $\triangle ABC$ inscribed in $\angle BAC$.

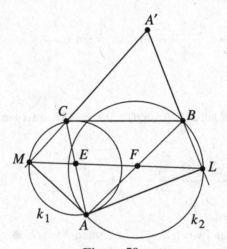

Figure 59.

First, notice that for any choice of A' the point M lies on the circle k_1 with diameter AC, while L lies on the circle k_2 with diameter AB (Fig. 59). Then obviously LM is maximal when the segment LM contains the centers E and F of k_1 and k_2. In this case we have

$$LM = AF + FE + EA = \frac{a+b+c}{2} = s.$$

Moreover, it follows from $\angle MEC = \angle AEF = \gamma$ that $\angle MCE = \angle CME = 90° - \frac{\gamma}{2}$, which in turn implies that MC is the bisector of the complementary angle to $\angle ACB$. The line LB has a similar property. Therefore A' is the center of the excircle for $\triangle ABC$ inscribed in $\angle BAC$. ◆

Problem 2.2.4 *For any point P in the plane different from the vertices A, B, and C of a given triangle ABC, set $x = AP$, $y = BP$, $z = CP$, $\alpha_1 = \angle BPC$, $\beta_1 = \angle APC$, and $\gamma_1 = \angle APB$. Find the position of P such that the sum*

$$q(P) = x \sin \alpha_1 + y \sin \beta_1 + z \sin \gamma_1$$

is maximal.

Solution. Denote by k the circumcircle of $\triangle BPC$ and by A' the intersection point of k and the line AP such that A' and P are on different sides of the line BC (Fig. 60). Let L and M be the feet of the perpendiculars drawn from A to the lines $A'B$ and $A'C$. We will show that $q(P) = LM$.

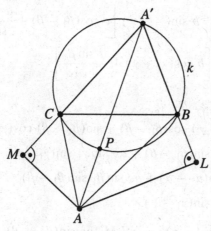

Figure 60.

Notice that $\angle PCB = \angle PA'B = \angle AML$ and $\angle PBC = \angle PA'C = \angle ALM$. Hence $\triangle PBC \sim \triangle ALM$, and therefore $\frac{a}{z} = \frac{LM}{AM}$. On the other hand, $\angle BCA' = \angle BPA' = 180° - \angle APB = 180° - \gamma_1$. Thus, $\angle ACM = 180° - \gamma - (180° - \gamma_1) = \gamma_1 - \gamma$, and so $AM = b\sin(\gamma_1 - \gamma)$. This implies

$$z = \frac{a\,AM}{LM} = \frac{ab\sin(\gamma_1 - \gamma)}{LM}.$$

Similarly, $x = \frac{bc\sin(\alpha_1 - \alpha)}{LM}$ and $y = \frac{ac\sin(\beta_1 - \beta)}{LM}$.

Next, Ptolemy's theorem for $ALA'M$ gives

$$AA' \cdot LM = AM \cdot A'L + A'M \cdot AL,$$

while the law of sines for $A'BC$ yields $A'B = \frac{a\sin\gamma_1}{\sin\alpha_1}$ and $A'C = \frac{a\sin\beta_1}{\sin\alpha_1}$. Since AA' is a diameter of the circumcircle of $ALA'M$, we have

$$\frac{LM}{\sin\alpha_1} = \frac{LM}{\sin\angle MAL} = AA'.$$

Also notice that

$$A'L = A'B + BL = \frac{a\sin\gamma_1}{\sin\alpha_1} + c\cos(\beta_1 - \beta)$$

and

$$A'M = A'C + CM = \frac{a \, \sin \beta_1}{\sin \alpha_1} + b \, \sin(\gamma_1 - \gamma).$$

Now the identity above implies

$$\frac{LM^2}{\sin \alpha_1} = AA' \cdot LM = b \, \sin(\gamma_1 - \gamma) \left[c \, \cos(\beta_1 - \beta) + \frac{a \, \sin \gamma_1}{\sin \alpha_1} \right]$$

$$+ c \, \sin \beta_1 \left[b \, \cos(\gamma_1 - \gamma) + \frac{a \, \sin \beta_1}{\sin \alpha_1} \right].$$

Therefore

$$LM^2 = bc \, [\sin(\gamma_1 - \gamma) \, \cos(\beta_1 - \beta) + \sin(\beta_1 - \beta) \, \cos(\gamma_1 - \gamma)]$$

$$+ ac \, \sin \beta_1 \, \sin(\beta_1 - \beta) + ab \, \sin \gamma_1 \, \sin(\gamma_1 - \gamma)$$

$$= bc \, \sin \alpha_1 \, \sin(\alpha_1 - \alpha) + ac \, \sin \beta_1 \, \sin(\beta_1 - \beta)$$

$$+ ab \, \sin \gamma_1 \, \sin(\gamma_1 - \gamma).$$

On the other hand, $bc \, \sin(\alpha_1 - \alpha) = x \, LM$, $ac \, \sin(\beta_1 - \beta) = y \, LM$, and $ab \, \sin(\gamma_1 - \gamma) = z \, LM$. Using these in the above equality for LM^2, one gets

$$LM = x \, \sin \alpha_1 + y \, \sin \beta_1 + z \, \sin \gamma_1 = q(P).$$

It now follows from Problem 2.2.3 that the sum $q(P)$ is maximal when A' is the center of the corresponding excircle. In this case we have $\gamma_1 = 90° + \frac{\gamma}{2}$, $\angle BAP = \frac{\alpha}{2}$, and therefore $\angle ABP = \frac{\beta}{2}$. Thus P is the incenter of $\triangle ABC$. ♠

EXERCISES

2.2.5 Given a point X in the interior of an acute triangle ABC, denote by A_1, B_1, and C_1 the intersection points of the lines AX, BX, and CX with the corresponding sides of the triangle. Show that the perimeter of triangle $A_1 B_1 C_1$ is minimal when X is the orthocenter of triangle ABC.

2.2.6 Given a point X in the interior of an acute triangle ABC, one draws the lines through X parallel to the sides of the triangle. These lines intersect the sides of the triangle at the points $M \in AC, N \in BC(MN||AB), P \in AB, Q \in AC(PQ||BC)$, and $R \in BC, S \in AB(RS||AC)$. Find the position of X such that the sum

$$MX \cdot NX + PX \cdot QX + RX \cdot SX$$

is a maximum.

2.2.7 Find the position of a point M inside an acute triangle ABC such that the sum:

(a) $AM \cdot BC + BM \cdot AC + CM \cdot AB$;

(b) $AM \cdot BM \cdot AB + BM \cdot CM \cdot BC + CM \cdot AM \cdot CA$

is a minimum.

2.2.8 Given a triangle ABC, find the points M in the plane such that the sum

$$AB \cdot MC^2 + BC \cdot MA^2 + CA \cdot MB^2$$

is a minimum.

2.2.9 Let M be a point in the interior of a triangle ABC and let A', B', and C' be the feet of the perpendiculars drawn from M to the lines BC, CA, and AB, respectively. Find the position of M such that

$$\frac{MA' \cdot MB' \cdot MC'}{MA \cdot MB \cdot MC}$$

is maximal.

2.2.10 For any point X in the interior of a triangle ABC set $m(X) = \min\{AX, BX, CX\}$. Find the position of X such that $m(X)$ is a maximum.

2.2.11 Triangle MNP is circumscribed about a given triangle ABC in such a way that the points A, B, and C lie on NP, PM, and MN, respectively, and $\angle PAB = \angle MBC = \angle NCA = \varphi$. Find the values of φ such that the area of triangle MNP is a maximum.

2.2.12 Find a point X in the interior of a regular tetrahedron such that the tetrahedron with vertices the orthogonal projections of X on its faces has maximum volume.

2.2.13 For any point X in the interior of a given tetrahedron $ABCD$ denote by X_1, X_2, X_3, and X_4 the orthogonal projections of X on the planes BCD, ACD, ABD, and ABC, respectively, and by x_1, x_2, x_3, and x_4 the distances from X to these planes. Set $S_1 = [BCD]$, $S_2 = [ACD]$, $S_3 = [ABD]$, $S_4 = [ABC]$.

(a) Prove that there exists a unique point X such that

$$\frac{x_1}{S_1} = \frac{x_2}{S_2} = \frac{x_3}{S_3} = \frac{x_4}{S_4}.$$

Denote this point by L and call it *Lemoine's point* for the tetrahedron $ABCD$.

(b) Show that the sum $x_1^2 + x_2^2 + x_3^2 + x_4^2$ is minimal precisely when X coincides with L.

(c) Show that L is the centroid of the tetrahedron $L_1 L_2 L_3 L_4$.

2.2.14 Inscribe a tetrahedron in a given tetrahedron such that the sum of squares of its edges is a minimum.

2.2.15 Find the position of a point inside a regular tetrahedron such that the sum of distances from it to the six edges of the tetrahedron is a minimum.

2.3 Malfatti's Problems

In 1803 the Italian mathematician Gianfrancesco Malfatti posed the following problem [13]: *Given a right triangular prism of any sort of material, such as marble, how shall three circular cylinders of the same height as the prism and of the greatest possible volume be related to one another in the prism and leave over the least possible amount of material?* This is equivalent to the plane problem of cutting three circles from a given triangle so that the sum of their areas is maximized.

As noted in [7], Malfatti, and many others who considered the problem, assumed that the solution would be the three circles that are tangent to each other, while each circle is tangent to two sides of the triangle (Fig. 61).

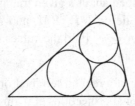

Figure 61.

These circles have become known as the Malfatti circles, and we refer the reader to [12] and [21] for some historical remarks on the derivation of their radii. In 1929, Lob and Richmond [12] noted that the Malfatti circles are not always the solution of the Malfatti problem. For example, in an equilateral triangle the in circle together with two little circles squeezed into the angles, contain a greater area than Malfatti's three circles. Moreover, Goldberg [7] proved in 1967 that the Malfatti circles never give a solution of the Malfatti problem. To the best of the authors' knowledge, the Malfatti problem was first solved by V. Zalgaller and G. Loss [23] in 1991.

They proved that for a triangle ABC with $\angle A \leq \angle B \leq \angle C$ the solution of the Malfatti problem is given by the circles k_1, k_2, k_3, where k_1 is the incircle, k_2 is inscribed in $\angle A$ and externally tangent to k_1, while k_3 is either the circle inscribed

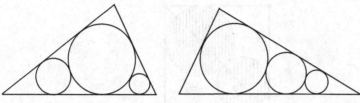

Figure 62.

in $\angle B$ and externally tangent to k_1 or the circle inscribed in $\angle A$ and externally tangent to k_2, depending on whether $\sin \frac{A}{2} \geq \tan \frac{B}{2}$ or $\sin \frac{A}{2} \leq \tan \frac{B}{2}$ (Fig. 62).

The proof of Zalgaller and Loss is very long (more than 25 pages) and we are not going to present it here. Instead, we shall consider the Malfatti problem for two circles in a square or a triangle, and we shall give a simple solution of the original Malfatti problem for an equilateral triangle.

We start with the Malfatti problem for two circles in a square.

Problem 2.3.1 *Cut two nonintersecting circles from a given square so that the sum of their areas is maximal.*

Solution. Assume that the side length of the square is 1, and consider two arbitrary nonintersecting circles inside of it (Fig. 63(a)). It is not difficult to see (the reader is advised to do this rigorously) that by moving the circles inside the square without intersecting them, they can be inscribed in opposite corners of the square (Fig. 63(b)).

Figure 63. (a)

Figure 63. (b)

Figure 63. (c)

Then one can increase the radius of one of them (which increases their total area) until they touch (Fig. 63(c)). Thus, it is enough to consider the case in which the two circles are situated as in Fig. 64.

If their radii are r_1 and r_2, then $\sqrt{2}\,r_1 + r_1 + r_2 + \sqrt{2}\,r_2 = \sqrt{2}$, so

$$r_1 + r_2 = 2 - \sqrt{2}.$$

Moreover, the fact that both circles lie entirely in the square implies

$$0 \leq r_1, r_2 \leq \frac{1}{2}.$$

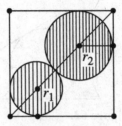

Figure 64.

Now the problem is to find the maximum value of the expression $r_1^2 + r_2^2$ under the conditions above. Assume for convenience that $r_1 \le r_2$. Then there exists x with $r_1 = \frac{2-\sqrt{2}}{2} - x$ and $r_2 = \frac{2-\sqrt{2}}{2} + x$, where $0 \le x \le \frac{\sqrt{2}-1}{2}$. Therefore $r_1^2 + r_2^2 = \frac{(2-\sqrt{2})^2}{2} + 2x^2$ is a maximum when $x = \frac{\sqrt{2}-1}{2}$. In this case $r_1 = \frac{3}{2} - \sqrt{2}$ and $r_2 = \frac{1}{2}$. Thus, the solution of the problem is given by the incircle of the square and one of the circles inscribed in a corner of the square that is tangent to the incircle (Fig. 65). ♠

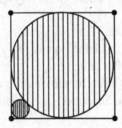

Figure 65.

We are now going to solve the more difficult Malfatti problem for two circles in a triangle.

Problem 2.3.2 *Cut two nonintersecting circles from a triangle such that the sum of their areas is maximal.*

Solution. Let k_1 and k_2 be two nonintersecting circles of radii r_1 and r_2 and centers O_1 and O_2 in triangle ABC. We may assume that each of them is tangent to at least two sides of the triangle. More specifically, assume that k_1 is tangent to AB and AC, while k_2 is tangent to AB and BC. Then O_1 and O_2 lie on the bisectors of angles A and B, respectively (Fig. 66). We may also assume that the two circles are tangent to each other; otherwise, enlarging one of them would clearly enlarge their total area.

Suppose now that neither k_1 nor k_2 coincides with the incircle k of $\triangle ABC$. We shall show that then there exists a circle k' of radius r' without common interior

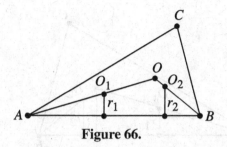

Figure 66.

points with k such that the sum of the areas of k and k' is greater than the sum of the areas of k_1 and k_2.

Assume for convenience that $\angle A \leq \angle B$. Without loss of generality we may assume that $r_1 \leq r_2$. Indeed, if $r_1 > r_2$, denote by k_1' the circle of radius $r_1' = r_2$ that is tangent to AB and BC, and by k_2' the circle of radius $r_2' = r_1$ that is tangent to AC and AB. Then $r_1' \leq r_2'$ and the total area of k_1' and k_2' is the same as that of k_1 and k_2. Moreover, $\angle A \leq \angle B$ implies $O_2'M \geq O_2N$ (Fig. 67). Hence $O_1'O_2' \geq O_1O_2 \geq r_1 + r_2 = r_1' + r_2'$, i.e., k_1' and k_2' have no common interior points.

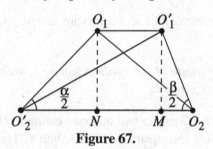

Figure 67.

So, we shall assume from now on that $\angle A \leq \angle B$ and $r_1 \leq r_2$. Set $\epsilon = r - r_2 > 0$, where r is the inradius of $\triangle ABC$. If $r_1 \leq \epsilon$, then $r_1 + r_2 \leq r$. Hence $r_1^2 + r_2^2 < r^2$, which means that the total area of k_1 and k_2 is less than the area of k. Now consider the case $r_1 > \epsilon$. Set $r' = r_1 - \epsilon$ and let k' be the circle of radius r' inscribed in $\angle A$ (Fig. 68). Then

$$r^2 + (r')^2 = (r_2 + \epsilon)^2 + (r_1 - \epsilon)^2$$
$$= r_1^2 + r_2^2 + 2\epsilon(r_2 - r_1) + 2\epsilon^2 > r_1^2 + r_2^2,$$

and it remains to show that k and k' have no common interior points. To do this we first note that

$$OO_2 = \frac{\epsilon}{\sin \frac{B}{2}} \leq \frac{\epsilon}{\sin \frac{A}{2}} = O_1O'.$$

Hence the triangle inequality implies that

$$OO' = OO_1 + O_1O' \geq OO_1 + O_2O \geq O_1O_2.$$

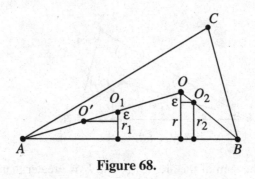

Figure 68.

Thus, $OO' \geq O_1O_2 \geq r_1 + r_2 = r + r'$ and therefore k and k' have no common interior points.

The above arguments show that two nonintersecting circles in $\triangle ABC$ have maximal combined area precisely when one of them is the incircle of $\triangle ABC$, while the other one is inscribed in the smallest angle of the triangle and is tangent to the incircle. ♠

Next, we consider two problems that in a sense are inverse to Problems 2.3.1 and 2.3.2.

Problem 2.3.3 *Find the side length of the smallest square containing two nonintersecting circles of given radii a and b.*

Solution. Assume that $a \geq b$. Consider two nonintersecting circles k_1 and k_2 of radii a and b, respectively, lying in a square S of side length x. Then the center O_1 (resp. O_2) of k_1 (resp. k_2) lies in the square whose sides are at distances a (resp. b) from the corresponding sides of S (Fig. 69). Then $O_1O_2 \leq AB = \sqrt{2}(x - a - b)$. On the other hand, $O_1O_2 \geq a + b$, since k_1 and k_2 do not intersect, and we get $\sqrt{2}(x - a - b) \geq a + b$. Hence

$$x \geq (a + b)\left(1 + \frac{1}{\sqrt{2}}\right).$$

It is clear also that $x \geq 2a$, since the circle k_1 lies inside the square S.

If $(a + b)\left(1 + \frac{1}{\sqrt{2}}\right) \geq 2a$, then the required smallest square has side of length $d = (a + b)\left(1 + \frac{1}{\sqrt{2}}\right)$. This follows from the inequalities above and the fact that in this case the two circles of radii a and b centered at A and B (see Fig. 69) are nonintersecting and lie in S.

Similarly, if $(a + b)\left(1 + \frac{1}{\sqrt{2}}\right) < 2a$, then the required smallest square has side of length $d = 2a$.

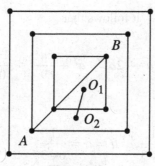

Figure 69.

Hence the solution of the problem is given by the square of side length

$$d = \begin{cases} (a+b)\left(1+\frac{1}{\sqrt{2}}\right) & \text{if } b(\sqrt{2}+1)^2 \geq a \geq b, \\ 2a & \text{if } a \geq b(\sqrt{2}+1)^2. \end{cases} \spadesuit$$

Using the same reasoning as above one can solve the analogous problem for an equilateral triangle.

Problem 2.3.4 *Show that the side length of the smallest equilateral triangle containing two nonintersecting circles of given radii a and b, a ≥ b, is given by*

$$d = \begin{cases} \sqrt{3}\,(a+b) + 2\sqrt{ab} & \text{if } b \leq a \leq 3b, \\ 2\sqrt{3}\,a & \text{if } a \geq 3b. \end{cases}$$

Now we shall use Problem 2.3.4 to solve the original Malfatti problem for an equilateral triangle.

Problem 2.3.5 *Prove that the solution of the Malfatti problem for an equilateral triangle is given by the incircle and two circles inscribed in its angles and tangent to the incircle.*

Solution. We may assume that the side length of the triangle is 1. Suppose that it contains three nonintersecting circles of radii $a \geq b \geq c$. Since the three circles from the statement of the problem have radii $\frac{1}{2\sqrt{3}}$, $\frac{1}{6\sqrt{3}}$, and $\frac{1}{6\sqrt{3}}$, we have to prove the following inequality:

(1) $$a^2 + b^2 + c^2 \leq \frac{11}{108}.$$

To do this we shall consider two cases.

Case 1 Let $a \geq 3b$. Since $a \leq \frac{1}{2\sqrt{3}}$, it follows that

$$a^2 + b^2 + c^2 \leq a^2 + 2b^2 \leq a^2 + \frac{2a^2}{9} \leq \frac{11}{108}.$$

The equality occurs if and only if

$$a = \frac{1}{2\sqrt{3}}, \quad b = c = \frac{1}{6\sqrt{3}}.$$

Case 2 Let $b \leq a \leq 3b$. Then it follows from Problem 2.3.4 that

$$\sqrt{3}\,(a + b) + 2\sqrt{ab} \leq 1.$$

Set $a = 3x^2 b$, where $x > 0$. Then the above inequalities are equivalent to

$$\frac{1}{\sqrt{3}} \leq x \leq 1, \quad b \leq \frac{1}{\sqrt{3}(3x^2 + 2x + 1)}.$$

Hence

$$a^2 + b^2 + c^2 \leq a^2 + 2b^2 = (9x^4 + 2)\,b^2 \leq \frac{9x^4 + 2}{3(3x^2 + 2x + 1)^2},$$

and it is enough to prove that

$$\frac{9x^4 + 2}{(3x^2 + 2x + 1)^2} \leq \frac{11}{36}$$

if $\frac{1}{\sqrt{3}} \leq x \leq 1$. The above inequality is equivalent to

$$(225\,x^3 + 93\,x^2 - 17\,x - 61)(x - 1) \leq 0,$$

which is satisfied since $x - 1 \leq 0$ and

$$225\,x^3 + 93\,x^2 - 17\,x - 61 = 51x(x^2 - 1/3) + 174\,x^3 + 93\,x^2 - 61$$

$$\geq \frac{174}{3\sqrt{3}} + \frac{93}{3} - 61 = \frac{174 - 90\sqrt{3}}{3\sqrt{3}} > 0.$$

In this case the equality in (1) is attained if and only if $x = 1$, $b = c = \frac{1}{\sqrt{3}(3x^2+2x+1)}$, giving again that $a = \frac{1}{2\sqrt{3}}$ and $b = c = \frac{1}{6\sqrt{3}}$. ♠

EXERCISES

2.3.6 Find the radii of the Malfatti circles for an equilateral triangle and show that they do not provide a solution to the Malfatti problem. Do the same for other types of triangles.

2.3.7 Cut two nonintersecting circles of radii r_1 and r_2 from a given square so that:

(a) $r_1 r_2$ is a maximum;   (b) $r_1^3 + r_2^3$ is a maximum.

2.3.8 Cut two nonintersecting circles from a given triangle such that the product of their areas is a maximum.

2.3.9 Cut two nonintersecting circles from a given rectangle such that:

(a) the sum of their areas;   (b) the product of their areas,

is a maximum.

2.3.10 Find the side length of the smallest square containing two nonintersecting circles of radii $\sqrt{2}$ and 2.

2.3.11 Find the side length of the smallest square containing three nonintersecting circles of radii 1, $\sqrt{2}$, and 2.

2.3.12 Find the side length of the smallest equilateral triangle containing three nonintersecting circles of radii 2, 3, and 4.

2.3.13 Solve the Malfatti problem for three circles in a square.

2.3.14 Find the side length of the smallest square containing 5 nonintersecting unit circles.

2.3.15 Cut two nonintersecting balls from a given cube such that:

(a) the sum of their volumes;   (b) the sum of their surface areas

is a maximum.

2.3.16 Find the edge length of the smallest cube containing two nonintersecting balls of given radii a and b.

2.3.17 Find the edge length of the smallest cube containing 9 nonintersecting unit balls.

2.4   Extremal Combinatorial Geometry Problems

The problems in the previous sections dealt with maxima and minima of geometric quantities like perimeter, area, volume, length of a segment, and measure of an angle. In this section we consider problems of a rather different nature. Namely, in most of them we will be concerned with the maximal or minimal number of points or figures in the plane (solids in space) having certain geometric properties.

Problem 2.4.1 *In a regular 2n-gon the midpoints of all its sides and diagonals are marked. What is the maximum number of marked points that lie on a circle?*

Solution. Let $A_1, A_2, \ldots, A_{2n}$ be the successive vertices of a regular $2n$-gon M, and let O be its center. For each $i = 1, 2, \ldots, n$ the midpoints of the diagonals (or sides) of M with length $A_1 A_{i+1}$ lie on a circle k_i with center O. Clearly k_i contains at most $2n$ points. Moreover, k_1 contains exactly $2n$ points, while $k_n = \{O\}$ (Fig. 70).

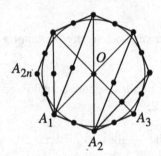

Figure 70.

Let us now show that every circle with center different from O contains fewer than $2n$ marked points. Indeed, if k is such a circle, then for any $i = 1, 2, \ldots, n-1$ it has at most 2 common marked points with k_i. Thus k contains at most $1 + 2(n - 1) = 2n - 1$ marked points.

Hence the maximum number of marked points on a circle is $2n$. ♠

Problem 2.4.2 *Given a coordinate system in the plane and an integer $n \geq 4$, find the maximum number of integer points (i.e., points with integer coordinates) that can be covered by a square of side length n.*

Solution. Consider an arbitrary square K with side length n and let M be the smallest convex polygon containing the integer points in K. Then the area $[M]$ of M does not exceed n^2, and its perimeter does not exceed $4n$. Using Pick's formula (see the Glossary), we have $[M] = \frac{m}{2} + k - 1$, where k is the number of integer points in the interior of M, while m is the number of integer points on the boundary of M. Hence $\frac{m}{2} + k - 1 \leq n^2$. Since the distance between any two distinct integer points is at least 1, the perimeter of M is at least m. Hence $m \leq 4n$ and we get

$$m + k = \left(\frac{m}{2} + k - 1\right) + \frac{m}{2} + 1 \leq n^2 + 2n + 1 = (n + 1)^2.$$

Thus, the number $m + k$ of the integer points in K does not exceed $(n + 1)^2$.

On the other hand, it is clear that there exists a square with side n covering $(n + 1)^2$ integer points. ♠

Problem 2.4.3 *A city has the form of a square with side of length 5 km. Its streets divide it into suburbs all of which are squares with sides of length 200 m. What is the maximum area of a region in the city bounded by a closed curve of length 10 km that consists entirely of streets or parts of streets of the city?*

Solution. Let C be an arbitrary closed curve consisting of streets or parts of streets, and let Π be the smallest rectangle containing C. Clearly the sides of Π are streets or parts of streets of the city and the perimeter of Π is not larger than the length of C.

Moreover, the area bounded by C is not larger than the area of Π. Thus, it is enough to consider only closed curves C of rectangular shape (Fig. 71).

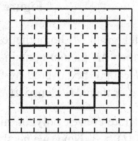

Figure 71.

Now consider a rectangle Π with perimeter 10 km whose boundary consists of streets or parts of streets. Denote by x the length of the smaller side of Π (in km). Then the length of the other side is $5 - x$, and $0 \leq x \leq \frac{5}{2}$. Moreover, $k = 5x$ is an integer with $0 \leq k \leq 12$. Thus, $[\Pi] = x(5 - x)$ is maximal when $x = \frac{5}{2}$. Moreover, the function $x(5 - x)$ is increasing for $x \in [0, 5/2]$, so for $x = \frac{k}{5}$ (with $k = 1, 2, \ldots, 12$) the maximum value of $[\Pi]$ is achieved when $x = \frac{12}{5}$. Hence the required closed curve must have the shape of a rectangle with sides $\frac{12}{5}$ km and $\frac{13}{5}$ km. ♠

Given a convex polygon M consider all homothetic images of M smaller than M. Denote by $n(M)$ the minimum number of such polygons that can cover M. As we will see in the next problem, the number $n(M)$ is the same for all polygons M that are not parallelograms. This remarkable fact is known as the *Gohberg–Markus theorem*. More details concerning this type of "covering problems" can be found in the book [3] of Boltyanskii and Gohberg.

Problem 2.4.4 *Let M be a convex nondegenerate (i.e., not lying on a line) polygon in the plane.*

(a) *If M is a parallelogram, then $n(M) = 4$.*

(b) *If M is not a parallelogram, then $n(M) = 3$.*

Solution.

(a) Let M be a parallelogram $ABCD$. It is easy to see that M can be covered by 4 smaller parallelograms homothetic to M (Fig. 72(a)).

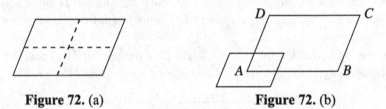

Figure 72. (a)   **Figure 72.** (b)

On the other hand, for any parallelogram M_1 homothetic to M and smaller than M, if M_1 contains the point A, then M_1 cannot contain any other vertex of M (Fig. 72(b)). This shows that M cannot be covered by fewer than 4 parallelograms homothetic to M and smaller than M. Hence $n(M) = 4$.

(b) Let M be an arbitrary nondegenerate convex polygon in the plane that is not a paralellogram. It is clear that $n(M) \geq 3$. Next, we need the following lemma.

Lemma. *There exists a triangle N containing M such that the line of every side of N contains a side of M.*

Proof of the Lemma. If M is a triangle, take $N = M$. Assume that M is not a triangle; then there exist two sides of M that are not parallel and have no common points. Extending these two sides until they intersect (Fig. 73(a)), one gets another convex polygon M_1 whose number of sides is less than that of M.

Continuing this process, after several steps one gets a parallelogram or a triangle M' containing M whose sides contain sides of M. If M' is triangle, set $N = M'$.

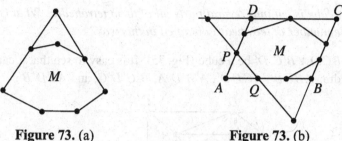

Figure 73. (a)   **Figure 73. (b)**

Assume that M' is a parallelogram $ABCD$. Since M is not a parallelogram, at least one of the vertices of M' is not a vertex of M. Suppose for example that A is not a vertex of M. Denote by P the point from M on the side AD that is closest to A and by Q the point from M on the side AB that is closest to A (Fig. 73(b)). Then the pentagon $QBCDP$ contains M. Moreover, the triangle N formed by the lines PQ, BC, and CD also contains M and has the required property. This proves the lemma. ♠

Using the lemma, consider a triangle ABC containing M and such that the sides A_1A_2, B_1B_2, and C_1C_2 of M lie on BC, AC, and AB, respectively (Fig. 74).

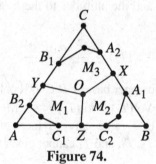

Figure 74.

Choose an arbitrary point O in the interior of M and arbitrary points X, Y, and Z in the interiors of the segments A_1A_2, B_1B_2, and C_1C_2, respectively (Fig. 74). Then the segments OX, OY, and OZ cut M into three polygons M_1, M_2, and M_3. Assume for example that M_1 is the polygon contained in the quadrilateral $AZOY$.

The choice of O, Y, and Z now shows that if $0 < k < 1$ and k is sufficiently close to 1, then the homothety φ_1 with center A and ratio k is such that $\varphi_1(M)$ contains $AZOY$ and therefore M_1. In the same way one derives that there exist homotheties φ_2 and φ_3 with coefficients less than 1 such that $\varphi_2(M)$ contains M_2, while $\varphi_3(M)$ contains M_3. Thus, M is contained in the union of $\varphi_1(M)$, $\varphi_2(M)$, and $\varphi_3(M)$, so $n(M) \leq 3$. This proves that $n(M) = 3$. ♠

To conclude this section we consider a space problem.

Problem 2.4.5 *A cube is cut into several parts all of them tetrahedra. What is the minimum possible number of tetrahedra obtained in this way?*

Solution. Let $ABCDA'B'C'D'$ be a cube (Fig. 75). It is easy to see that it can be cut into 5 tetrahedra: $ABCB'$, $ACDD'$, $A'B'D'A$, $B'C'D'C$, and $ACD'B'$.

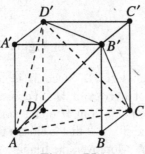

Figure 75.

We are now going to show that 5 is the desired number. Let $a = AB$. Assume that the cube is cut into several tetrahedra. Clearly the base $ABCD$ must contain faces of at least two different tetrahedra T_1 and T_2. If the areas of these two faces are S_1 and S_2, then $S_1 + S_2 \leq a^2$ and the altitudes to these faces in T_1 and T_2 are not longer than a. Hence

$$\text{Vol}(T_1) + \text{Vol}(T_2) \leq \frac{a^2 \cdot a}{3} = \frac{a^3}{3}.$$

In a similar way one shows that the upper base $A'B'C'D'$ contains the faces of two different tetrahedra T_3 and T_4 with $\text{Vol}(T_3)+\text{Vol}(T_4) \leq \frac{a^3}{3}$. Moreover, it is clear that T_1 and T_2 cannot coincide with T_3 or T_4, since any two faces of a tetrahedron have a common edge. The tetrahedra T_1, T_2, T_3, and T_4 cannot cover the whole cube since

$$\text{Vol}(T_1) + \text{Vol}(T_2) + \text{Vol}(T_3) + \text{Vol}(T_4) \leq 2\frac{a^3}{3} < a^3,$$

So, there must be at least one more tetrahedron obtained by the cutting of the cube. Hence if a cube is cut into tetrahedra, their number is at least 5. ♠

EXERCISES

Cuttings

2.4.6 What is the maximum number of triangles into which a given triangle ABC can be cut so that the number of segments meeting at any vertex of the net obtained in this way is the same and all vertices except A, B, and C lie in the interior of ABC.

2.4.7 Find the minimum number of planes required to cut a given cube into at least 300 pieces.

2.4.8 What is the minimum width of an infinite horizontal strip of the plane from which an arbitrary triangle of area 1 can be cut off.

MaxMin and MinMax

2.4.9 Let $n \geq 3$ be a given integer. For any points $A_1, A_2, \ldots, A_n$ in the plane no three of which lie on a line, denote by α the smallest of the angles $A_i A_j A_k$ for different i, j, and k. Find the largest possible value of α.

2.4.10 Let A_1, A_2, A_3, A_4 be arbitrary points on the boundary or in the interior of a given rectangle with sides of lengths 3 and 4. Prove that

$$\max(\min_{1 \leq i \neq j \leq 4} A_i A_j) = \frac{25}{8}.$$

2.4.11 Consider n arbitrary segments of length 1 in the plane, intersecting at one point. Show that the length of at least one side of the $2n$-gon with vertices the ends of the segments is not less than the side length of a regular $2n$-gon inscribed in a circle with diameter 1.

Angles

2.4.12 What is the maximal possible number of acute angles of a convex polygon?

2.4.13 Find the largest possible number of rays in space issuing from a point such that the angle between any two of them is:

(a) greater than 90°;   (b) greater than or equal to 90°.

2.4.14 Find the largest possible number of points

(a) in the plane;   (b) in space,

such that no triangle with vertices at these points has an obtuse angle.

Distribution of points

2.4.15 What is the largest number of points that can be distributed in a unit disk such that the distance between any two of them is greater than 1?

2.4.16 What is the least number of points that can be distributed in a convex n-gon such that every triangle with vertices at the vertices of the n-gon contains at least one of these points?

2.4.17 Find a rectangle T of minimum possible area such that for any position of T in the plane it contains a point with integer coordinates in its interior or on its boundary.

Chapter 3

Miscellaneous

3.1 Triangle Inequality

Problem 3.1.1 *Let X and Y be points on the sides AC and BC of an equilateral triangle ABC. Find the minimum and the maximum of the sum of orthogonal projections of the segment XY on the sides of ABC.*

Problem 3.1.2 *Find the least possible real number k for which the following statement is true: in every triangle one can find two sides of lengths a and b such that $1 \le \frac{a}{b} < k$.*

Problem 3.1.3 *Find the greatest real number k such that for any triple of positive numbers a, b, c such that $kabc > a^3 + b^3 + c^3$, there exists a triangle with side lengths a, b, c.*

Problem 3.1.4 *Let a, b, c be positive numbers such that*

$$abc \le \frac{1}{4} \text{ and } \frac{1}{a^2} + \frac{1}{b^2} + \frac{1}{c^2} < 9.$$

Prove that there exists a triangle with side lengths a, b, and c.

Problem 3.1.5 *Consider the inequality*

$$a^3 + b^3 + c^3 < k(a + b + c)(ab + bc + ca),$$

where a, b, c are the side lengths of a triangle and k is a real number.

(a) *Prove the inequality when $k = 1$.*

(b) *Find the least value of k such that the inequality holds true for any triangle.*

Problem 3.1.6 *Let* a, b, c *be positive real numbers. Prove that they are side lengths of a triangle if and only if*

$$a^2 pq + b^2 qr + c^2 rp < 0$$

for any real numbers p, q, r *such that* $p + q + r = 0$, $pqr \neq 0$.

Problem 3.1.7 *Let* x, y, z *be real numbers. Prove that the following conditions are equivalent:*

(i) $x, y, z > 0$ *and* $\frac{1}{x} + \frac{1}{y} + \frac{1}{z} \leq 1$.

(ii) $a^2 x + b^2 y + c^2 z > d^2$ *for every quadrilateral with side lengths* a, b, c, d.

3.2 Selected Geometric Inequalities

Problem 3.2.1 *Let* s, R, *and* r *be the semiperimeter, the circumradius, and the inradius of a triangle with side lengths* a, b, *and* c. *Prove that:*

(i) $(a + b - c)(b + c - a)(c + a - b) \leq abc$;

(ii) $R \geq 2r$ *(Euler's inequality)*;

(iii) $|s^2 - 2R^2 - 10Rr + r^2| \leq 2(R - 2r)\sqrt{R(R - 2r)}$ *(fundamental inequality)*;

(iv) $24Rr - 12r^2 \leq a^2 + b^2 + c^2 \leq 8R^2 + 4r^2$;

(v) $6\sqrt{3}r \leq a + b + c \leq 4R + (6\sqrt{3} - 8)r$.

Problem 3.2.2 *Let* M *and* N *be points on the sides* AC *and* BC *of a triangle* ABC *and let* L *be a point on the segment* MN. *Prove that*

$$\sqrt[3]{S} \geq \sqrt[3]{S_1} + \sqrt[3]{S_2},$$

where $S = [ABC]$, $S_1 = [AML]$, *and* $S_2 = [BNL]$.

Problem 3.2.3 *Let* M *be an interior point of a triangle* ABC *and* A', B', C' *its orthogonal projections on the lines* BC, CA, AB, *respectively. Prove that*

(i) $MA + MB + MC \geq 2(MA' + MB' + MC')$ *(Erdős–Mordell inequality)*.

(ii) $\frac{1}{MA} + \frac{1}{MB} + \frac{1}{MC} \leq \frac{1}{2}\left(\frac{1}{MA'} + \frac{1}{MB'} + \frac{1}{MC'}\right)$.

Problem 3.2.4 *Let ABC be a triangle inscribed in a circle of radius R, and let M be a point in the interior of ABC. Prove that*

$$\frac{MA}{BC^2} + \frac{MB}{CA^2} + \frac{MC}{AB^2} \geq \frac{1}{R}.$$

Problem 3.2.5 *Let $ABCDEF$ be a convex hexagon such that AB is parallel to DE, BC is parallel to EF, and CD is parallel to AF. Let R_A, R_C, R_E denote the circumradii of triangles FAB, BCD, DEF respectively, and let P denote the perimeter of the hexagon. Prove that*

$$R_A + R_C + R_E \geq \frac{P}{2}.$$

Problem 3.2.6 *Let A, B, C, and D be arbitrary points in the plane. Prove that*

$$AB.CD + AD.BC \geq AC.BD$$

(Ptolemy's inequality).

Problem 3.2.7 *Let $ABCDEF$ be a convex hexagon such that $AB = BC$, $CD = DE$, $EF = FA$. Prove that*

$$\frac{BC}{BE} + \frac{DE}{DA} + \frac{FA}{FC} \geq \frac{3}{2}.$$

When does equality occur?

Problem 3.2.8 *Let O be a point inside a convex quadrilateral $ABCD$ of area S and K, L, M, N interior points of the sides AB, BC, CD, DA, respectively, such that $OKBL$ and $OMDN$ are parallelograms. Prove that*

$$\sqrt{S} \geq \sqrt{S_1} + \sqrt{S_2},$$

where S_1 and S_2 are the areas of $ONAK$ and $OLCM$, respectively.

Problem 3.2.9 *A point O and a polygon F (not necessarily convex) are given in the plane. Let P denote the perimeter of F, D the sum of the distances from O to the vertices of F, and H the sum of the distances from O to the lines containing the sides of F. Prove that $D^2 - H^2 \geq \frac{P^2}{4}$.*

Problem 3.2.10 *Let $A_1, A_2, \ldots, A_{2n}, n \geq 2$, be arbitrary points in the plane. Denote by B_k, $1 \leq k \leq 2n$, the midpoint of the segment $A_k A_{k+1}$ $(A_{2n+1} = A_1)$. Prove that*

$$\sum_{k=1}^{n} (A_k A_{k+1} + A_{n+k} A_{n+k+1})^2 \geq 4 \tan^2 \frac{\pi}{2n} \sum_{k=1}^{n} B_k B_{k+n}^2.$$

Problem 3.2.11 *Let $C_1, C_2, C_3, \ldots, C_n, n \geq 3$, be unit circles in the plane, with centers $O_1, O_2, O_3, \ldots, O_n$, respectively. If no line meets more than two of the circles, prove that*

$$\sum_{1 \leq i < j \leq n} \frac{1}{O_i O_j} \leq \frac{(n-1)\pi}{4}.$$

3.3   MaxMin and MinMax

Problem 3.3.1 *Given a trapezoid of area 1, find the least possible length of its longest diagonal.*

Problem 3.3.2 *In triangle ABC, $\angle C = 90°, \angle A = 30°$, and $BC = 1$. Find the minimum of the length of the longest side of a triangle inscribed in ABC (that is, one such that each side of ABC contains a different vertex of the triangle).*

Problem 3.3.3 *For which acute-angled triangle is the ratio of the shortest side to the inradius maximal?*

Problem 3.3.4 *For any five points in the plane, denote by λ the ratio of the greatest distance to the smallest distance between two of them.*

(a) *Prove that $\lambda \geq 2 \sin 54°$.*

(b) *Determine when equality holds.*

Problem 3.3.5 *Let C be a unit circle and n a fixed positive integer. For any set A of n points $P_1, P_2, \ldots, P_n$ on C define*

$$D(A) = \max_d (\min_i \delta(P_i, d)),$$

where $\delta(P, l))$ denotes the distance from point P to line l and the maximum is taken over all diameters d of circle C. Let $\mathcal{F}_n$ be the family of all n-element subsets $A \subset C$ and let

$$D_n = \min_{A \in \mathcal{F}_n} D(A).$$

Calculate D_n and describe all sets $A \in \mathcal{F}_n$ with $D(A) = D_n$.

3.4 Area and Perimeter

Problem 3.4.1 *Let the points P and Q on AB, and R on AC, divide the perimeter of triangle ABC into three equal parts. Prove that the area of triangle PQR is greater than $\frac{2}{9}$ the area of triangle ABC.*

Problem 3.4.2 *In triangle ABC, angle A is twice angle B, angle C is obtuse, and the three sides have integer lengths. Determine the minimum possible perimeter of the triangle.*

Problem 3.4.3 *Prove that the area of a triangle with vertices on the sides of a parallelogram is not greater than one-half the area of the parallelogram.*

Problem 3.4.4 *A parallelogram of area S lies inside a triangle of area T. Prove that $T \geq 2S$.*

Problem 3.4.5 *Let PQRS be a convex quadrilateral inside a triangle ABC. Prove that the area of one of triangles PQR, PQS, PRS, and QRS is not greater than $\frac{1}{4}$ the area of triangle ABC.*

Problem 3.4.6 *Find a centrally symmetric polygon of maximal area contained in a given triangle.*

Problem 3.4.7 *Two equilateral triangles are inscribed in a circle with radius r. Let K be the area of the set consisting of all points interior to both triangles. Find the minimum of K.*

Problem 3.4.8 *Find the maximum possible value of the inradius of a triangle with vertices on the boundary or in the interior of a unit square.*

Problem 3.4.9 *Given a positive integer n cut n rectangles from an acute triangle ABC such that all of them have a side parallel to AB and their total area is a maximum.*

Problem 3.4.10 *The octagon $P_1P_2P_3P_4P_5P_6P_7P_8$ is inscribed in a circle, with the vertices around the circumference in the given order. Given that the polygon $P_1P_3P_5P_7$ is a square of area 5, and the polygon $P_2P_4P_6P_8$ is a rectangle of area 4, find the maximum possible area of the octagon.*

Problem 3.4.11 *Given a trapezoid ABCD (AB||CD) and a point K on ˙AB, find the point M on CD such that the area of the common part of triangles ABM and CDK is maximized.*

Problem 3.4.12 *Let ABC be a triangle. Prove that there is a line l (in the plane of triangle ABC) such that the intersection of the interior of triangle ABC and the interior of its reflection $A'B'C'$ in l has area more than 2/3 the area of triangle ABC.*

Problem 3.4.13 *To clip a convex n-gon means to choose a pair of consecutive sides AB, BC and to replace them by the segments AM, MN, and NC, where M is the midpoint of AB and N is the midpoint of BC. In other words, one cuts the triangle MBN to obtain a convex $(n+1)$-gon. A regular hexagon P_6 of area 1 is clipped to obtain a heptagon P_7. Then P_7 is clipped (in one of the seven possible ways) to obtain an octagon P_8, and so on. Prove that no matter how the clippings are done, the area of P_n is greater than 1/3, for all $n > 6$.*

Problem 3.4.14 *Prove that any convex pentagon whose vertices have integer co-ordinates must have area greater than or equal to $\frac{5}{2}$.*

Problem 3.4.15 *Each side of a convex polygon has integral length and the perimeter is odd. Prove that the area of the polygon is at least $\frac{\sqrt{3}}{4}$.*

Problem 3.4.16 *Let the area and the perimeter of a cyclic quadrilateral C be A_C and P_C, respectively. If the area and the perimeter of the quadrilateral that is tangent to the circumcircle of C at the vertices of C are A_T and P_T, respectively, prove that*

$$\frac{A_C}{A_T} \geq \left(\frac{P_C}{P_T}\right)^2.$$

Problem 3.4.17 *Two concentric circles have radii r and R respectively, where $R > r$. A convex quadrilateral ABCD is inscribed in the smaller circle and the extensions of AB, BC, CD, and DA intersect the larger circle at C_1, D_1, A_1, and B_1, respectively. Prove that:*

(a) *The perimeter of $A_1B_1C_1D_1$ is not less than $\frac{R}{r}$, the perimeter of ABCD.*

(b) *The area of $A_1B_1C_1D_1$ is not less than $\left(\frac{R}{r}\right)^2$, the area of ABCD.*

Problem 3.4.18 *An infinite square grid is colored in the chessboard pattern. For any pair of positive integers m, n consider a right-angled triangle whose vertices are grid points and whose legs, of length m and n, go along the lines of the grid. Let S_b be the total area of the black part of the triangle and S_w the total area of its white part. Define the function $f(m,n) = |S_b - S_w|$.*

(a) *Calculate $f(m, n)$ for all numbers m, n that have the same parity.*

(b) *Prove that $f(m, n) \leq \frac{1}{2} \max(m, n)$.*

(c) *Show that $f(m, n)$ is not bounded from above.*

3.5 Polygons in a Square

Problem 3.5.1 *A triangle of area $\frac{1}{2}$ lies in a unit square. Prove that at least two of its vertices are vertices of the square.*

Problem 3.5.2 *A quadrilateral is inscribed in a unit square. Prove that at least one of its sides has length not less than $\frac{\sqrt{2}}{2}$.*

Problem 3.5.3 *Find the minimum and the maximum of the area of an equilateral triangle inscribed in a unit square.*

Problem 3.5.4 *A convex polygon of area greater than $\frac{1}{2}$ lies in a unit square. Prove that the polygon contains a line segment of length $\frac{1}{2}$ that is parallel to a side of the square.*

Problem 3.5.5 *A convex n-gon lies in a unit square. Show that three of its vertices form a triangle of area less than:*

(a) $\frac{8}{n^2}$;

(b) $\frac{8}{n^2} \sin \frac{2\pi}{n}$.

Problem 3.5.6 *In a unit square a finite number of line segments parallel to its sides are drawn. The line segments may intersect one another and their total length is 18. Prove that at least one of the regions into which the square is divided by the line segments has area not less than 0.01.*

3.6 Broken Lines

Problem 3.6.1 *A broken line of length l is drawn in a unit square so that any line parallel to a side of the square intersects it at most once. Prove that:*

(a) $l < 2$;

(b) *for any $l \in (0, 2)$, there is a broken line of length l with the given property.*

Problem 3.6.2 *Two broken lines are given such that the distance between any two vertices of one broken line is at most 1, but the distance between any two vertices of different broken lines is more than $\frac{1}{\sqrt{2}}$. Prove that the broken lines have no common point.*

Problem 3.6.3 *An ant crosses a circular disk of radius r and it advances in a straight line, but sometimes it stops. Whenever it stops, it turns $60°$, each time in the opposite direction. (If the last time it turned $60°$ clockwise, this time it turns $60°$ counterclockwise.) Find the maximum length of the ant's path.*

Problem 3.6.4 *A non-self-intersecting broken line of length 1000 is drawn in a unit square. Prove that there exists a line parallel to a side of the square and intersecting the broken line at least 500 times.*

Problem 3.6.5 *Consider n^2 arbitrary points in a unit square. Show that there exists a broken line with vertices at these points whose length is not greater than $2n$.*

Problem 3.6.6 *A country with the shape of a square of side length 1000 km has 51 towns. Its government has an amount of money to construct highways of total length 11000 km. Is that amount of money enough to construct a system of highways connecting all towns of the country?*

Problem 3.6.7 *A broken line of length l is drawn in a unit square so that any point of the square is at distance less than d from a point of the broken line. Prove that $l \geq \frac{1}{2d} - \frac{\pi d}{2}$.*

3.7   Distribution of Points

Problem 3.7.1 *Let S be a set of finitely many points on the sides of a unit square. Prove that there is a vertex of the square such that the arithmetic mean of the squares of the distances from it to all points of S is not less than $\frac{3}{4}$.*

Problem 3.7.2 *Prove that among any 101 points in a unit square there are at least five lying in a circle of radius $\frac{1}{7}$.*

Problem 3.7.3 *Prove that among any 112 points in a unit square there are two at distance less than $\frac{1}{8}$.*

Problem 3.7.4 *Eight points are given in the interior or on the boundary of a unit cube such that any two of them are at least distance 1 apart. Prove that these points coincide with the vertices of the cube.*

Problem 3.7.5 *In a square of side length 100 there are given n nonintersecting unit disks such that any line segment of length 10 has a common point with at least one of them. Prove that $n \geq 400$.*

Problem 3.7.6 *Given n points inside a unit square, prove that:*

(a) *the area of at least one triangle with vertices at the given points or the vertices of the square is not greater than $\frac{1}{2(n+1)}$;*

(b) *the area of at least one triangle with vertices at the given n points ($n \geq 3$) is less than $\frac{1}{n-2}$.*

Problem 3.7.7 *Let $P_i(x_i, y_i)$, $1 \leq i \leq 6$, be points in the plane such that $x_i = 0, \pm 1$, or ± 2 and $y_i = 0, \pm 1$, or ± 2. Moreover, no three of these six points are collinear. Prove that there exists a triangle $P_i P_j P_k$, $1 \leq i < j < k \leq 6$, that has area not greater than 2.*

Problem 3.7.8 *Let S be a set of 1980 points in the plane. Every two points of S are at least distance 1 apart. Prove that S contains a subset T of 248 points, every two at least distance $\sqrt{3}$ apart.*

Problem 3.7.9 *In an annulus determined by two concentric circles of radii 1 and $\sqrt{2}$, respectively, there are given n points such that the distance between any two of them is not less than 1. Find the largest n for which this is possible.*

Problem 3.7.10 *Ten gangsters are standing on a flat surface, and the distances between them are all distinct. At twelve o'clock, when the church bells start chiming, each of them shoots at the one among the other nine gangsters who is the nearest and kills him or her. At least how many gangsters will be killed?*

Problem 3.7.11 *In a plane a set of n points ($n \geq 3$) is given. Each pair of points is connected by a segment. Let d be the length of the longest of these segments. We define a diameter of the set to be any connecting segment of length d. Prove that the number of diameters of the given set is at most n.*

Problem 3.7.12 *Given $n > 4$ points in the plane such that no three are collinear, prove that there are at least $\binom{n-3}{2}$ convex quadrilaterals whose vertices are four of the given points.*

3.8 Coverings

Problem 3.8.1 *The lengths of all sides and both diagonals of a quadrilateral are less than 1. Prove that it may be covered by a circle of radius $\frac{1}{\sqrt{3}}$.*

Problem 3.8.2 *Let $ABCD$ be a parallelogram with side lengths $AB = a$, $AD = 1$ and with $\angle BAD = \alpha$. If triangle ABD is acute, prove that the four circles of radius 1 with centers A, B, C, D cover the parallelogram if and only if $a \leq \cos \alpha + \sqrt{3} \sin \alpha$.*

Problem 3.8.3 *An equilateral triangle of side length 1 is covered by six congruent circles of radius r. Prove that $r \geq \frac{1}{4}(\sqrt{3} - 1)$.*

Problem 3.8.4 *Find the side length of the largest equilateral triangle that can be covered by three equilateral triangles of side lengths 1.*

Problem 3.8.5 *Find the radius of the largest disk that can be covered by:*

(a) *three unit disks;*

(b) *three disks with radii R_1, R_2, and R_3.*

Problem 3.8.6 *Find the minimum number of unit disks that can cover a disk of radius 2.*

Problem 3.8.7 *Is it possible to cover a square of side length $\frac{5}{4}$ by means of three unit squares?*

Problem 3.8.8 *Show that one can cover a unit square by means of any finite collection of squares of total area 4.*

Chapter 4

Hints and Solutions to the Exercises

4.1 Employing Geometric Transformations

1.1.11 Let C' be the symmetric point of C with respect to M (Fig. 76).

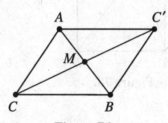

Figure 76.

Then $C'A = CB$, and the triangle inequality gives

$$CM = \frac{1}{2}CC' \le \frac{1}{2}(CA + C'A) = \frac{1}{2}(CA + CB).$$

1.1.12 Let C' be the point such that $ADCC'$ is a parallelogram and C'' the midpoint of $C'B$ (Fig. 77). Then $C''N = \frac{1}{2}C'C = \frac{1}{2}AD$ and $C''N \parallel CC' \parallel AD$. Hence $AC''NM$ is a parallelogram, implying $MN = AC''$. Now it follows from Problem 1.1.11 that

$$MN = AC'' \le \frac{1}{2}(AB + AC') = \frac{1}{2}(AB + CD).$$

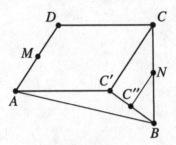

Figure 77.

1.1.13 Let a be the side length of the given square $ABCD$, and let X be an arbitrary point on its boundary, say on the side CD (Fig. 78). Then $s(X) = XA + XB + XC + XD = XA + XB + CD$. Heron's problem (Problem 1.1.1) implies that $AX + BX$ is minimal when $\angle AXD = \angle BXC$, i.e., when X is the midpoint of CD.

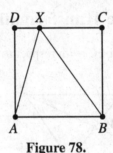

Figure 78.

In this case $s(X) = (\sqrt{5} + 1)a$. The minimum value of $s(X)$ is obtained also when X is the midpoint of AB, BC, or AD.

1.1.14 *Hint.* One may assume that two of the vertices of the triangles considered are fixed, while the third vertex lies on a fixed line parallel to the line determined by the first two vertices. Then one can use the argument from the solution of Problem 1.1.1.

1.1.15 Let B' be the reflection of B in ℓ. If $B' = A$, then $AX - BX = 0$ for any point X on ℓ. Assume that $B' \neq A$ and that the line AB' intersects ℓ at some point X_0 (Fig. 79). Then the triangle inequality implies $|AX - BX| = |AX - XB'| \leq AB'$ for every point X on ℓ, where equality holds only when $X = X_0$. Hence in this case the solution is given by the point X_0.

We leave as in exersice to the reader to show that if $B' \neq A$ and $AB' \parallel \ell$, then $|AX - BX|$ has no maximum.

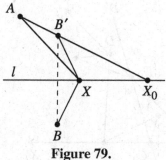

Figure 79.

1.1.16 Let P' and P'' be the reflections of P in the lines OX and OY, respectively. If $\angle XOY < 90°$, then $\angle P'OP'' = 2\angle XOY < 180°$ and therefore the line segment $P'P''$ intersects OX at point A_0 and OY at point B_0 (Fig. 80).

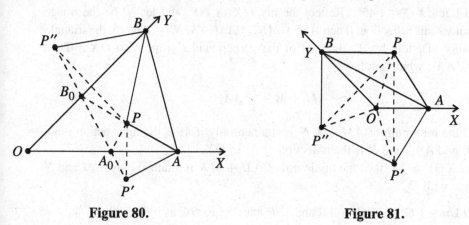

Figure 80.   **Figure 81.**

Given points A on OX and B on OY, the perimeter of triangle PAB is equal to the length of the broken line $P'ABP''$. Hence A_0 and B_0 are the desired points since in this case the perimeter of triangle PA_0B_0 is equal to the length of the line segment $P'P''$.

If $\angle XOY \geq 90°$, then $P'P''$ does not intersect the sides OX and OY of the angle, and the required points A and B coinside with O (Fig. 81).

1.1.17 Let A' be the reflection of the point A in the line OX and B' the reflection of B in OY (Fig. 82).

For any points C on OX and D on OY the length of the broken line $ACDB$ coincides with the length of the broken line $A'CDB'$. It is clear now that if $A'B'$ has no common points with the rays OX and OY, the required broken line is shortest when $C = D = O$. If $A'B'$ intersects OX at some point C_0, and OY at some point D_0, then the required broken line is AC_0D_0B.

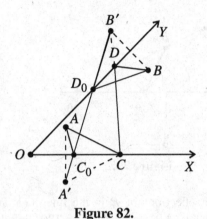

Figure 82.

1.1.18 Let $\angle XOY < 45°$. Reflect the ray OX in OY, and let N' be the image of N under this reflection. Then $AM + MN = AM + MN' \geq AN'$ by the triangle inequality. Denote by A_1 the foot of the perpendicular from A to OX'. Then $AN' \geq AA_1$, which implies

$$AM + MN \geq AA_1.$$

Hence the minimum of $AM + MN$ is attained only if M is the intersection point of OY and AA_1, and N is the reflection of A_1 in OY.

If $\angle XOY \geq 45°$ then the minimum of $AM + MN$ is attained only if M and N coincide with O.

1.1.19 Draw PC' parallel to AB and $C'P'$ parallel to BC as in Fig. 83.

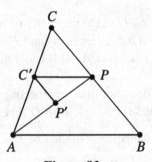

Figure 83.

Since $\triangle AC'P'$ is similar to $\triangle ACP$ and $\triangle PC'P'$ is similar to $\triangle ABP$, we have

$$\frac{C'P'}{CP} = \frac{AP'}{AP} \quad \text{and} \quad \frac{C'P'}{BP} = \frac{P'P}{AP}.$$

Adding up these equalities yields

$$\frac{C'P'}{BP} + \frac{C'P'}{CP} = \frac{P'P + AP'}{AP} = 1.$$

Therefore

$$\frac{1}{BP} + \frac{1}{CP} = \frac{1}{C'P'}.$$

Maximizing the expression on the left is then equivalent to minimizing $C'P'$. But C' does not depend on the choice of BC, so the latter reduces to finding a point P' on AP at minimum distance from C'. Clearly, this point is the foot of the perpendicular from C' to AP. Since $C'P'$ is parallel to BC by construction, the maximum of $1/BP + 1/CP$ is assumed only if BC is perpendicular to AP.

1.1.20 We shall prove that the desired line is parallel to BD. Indeed, denote by M_0 and K_0 the intersection points of this line with the lines AB and AD (Fig. 84). Then we have to prove that

$$\frac{1}{[BMC]} + \frac{1}{[DCK]} > \frac{1}{[BM_0C]} + \frac{1}{[DCK_0]},$$

which is equivalent to

$$(1) \qquad \frac{1}{[BMC]} - \frac{1}{[BM_0C]} > \frac{1}{[DCK_0]} - \frac{1}{[DCK]}.$$

We may assume that $M \in BM_0$. Then $K_0 \in DK$ and (1) is equivalent to

$$(2) \qquad \frac{[MCM_0]}{[BMC][BM_0C]} > \frac{[KCK_0]}{[DCK_0][DCK]}.$$

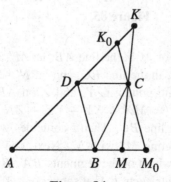

Figure 84.

Taking into account that $\angle MCM_0 = \angle KCK_0$, we see that (2) can be written as

(3)
$$\frac{MC}{[BMC]} \cdot \frac{CM_0}{[BM_0C]} > \frac{KC}{[DCK]} \cdot \frac{CK_0}{[DCK_0]}.$$

On the other hand,

$$\frac{CM_0}{[BM_0C]} = \frac{CK_0}{[DCK_0]}$$

since $M_0K_0 \| BD$ and (3) is equivalent to

$$\frac{MC}{[BMC]} > \frac{KC}{[DCK]}.$$

The latter inequality holds true since obviously the distance from B to MK is shorter than the distance from D to MK.

1.1.21 Let α be the measure of the given angle, and let M' be the image of M under rotation through α counterclockwise about O (Fig. 85). If A and B are points on OX and OY, respectively, and $OA = OB$, then $\triangle OAM \cong \triangle OBM'$, so $AM = BM'$. Hence $MA + MB = MB + BM' \geq MM'$. Thus $MA + MB$ is a minimum when B coincides with the intersection point of MM' and OY.

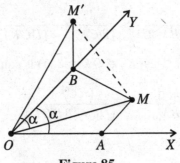

Figure 85.

1.1.22 Let M' be the reflection of M in the line AB, let M'' and A' be the reflections of M' and A, respectively, in the line BC, and let N' be the reflection of N in the line AC. We want to find points X, Y, and Z on AB, BC, and CA, respectively, such that the sum $t = MX + XY + YZ + ZN$ is a minimum. Let X' be the reflection of X in the line BC. Then t coincides with the length of the broken line $M''X'YZN'$ connecting M'' with N'. Next, one has to consider several possible cases concerning which of the segments BA', BC, and AC intersect $M''N'$. For example, if $M''N'$ intersects BA' at some point X'_0, BC at Y_0, and AC

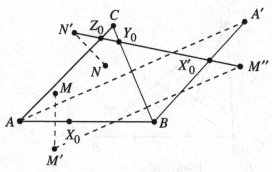

Figure 86.

at Z_0 (Fig. 86), then the required minimal path will be $MX_0Y_0Z_0N$, where X_0 is the reflection of X_0' in BC.

If $M''N'$ intersects BA' at X_0' and has no common point with BC and AC, we set $Y_0 = Z_0 = C$ and choose X_0 as above, etc.

1.1.23 *Hints.*

(a) Show that the minimum of the sum considered is attained when ℓ coincides with one of the lines AC and BC.

(b) Show that the required maximum is equal to $\max\{AB, AB'\}$ and it is attained when ℓ is perpendicular to AB or AB', where B' is the reflection of B in C.

1.1.24 By the minimum choice of M_ℓ, the inequality $AM_\ell + BM_\ell \le AC + BC$ holds true. Therefore the maximum value of $AM_\ell + BM_\ell$ does not exceed $AC + BC$. We prove that this maximum value is in fact equal to $AC + BC$. It suffices to construct a line ℓ through C such that $M_\ell = C$. We distinguish several cases.

If C is on the line segment AB, then $C = M_\ell$ for each line ℓ through C. If C lies on the line AB but not on the line segment AB, then it follows from Heron's problem that $C = M_\ell$ only for the line ℓ through C that is perpendicular to AB. (This is because if M is a point on ℓ different from C, then $CA < MA$ and $CB < MB$.)

Finally, suppose C is not on the line AB. Then the exterior bisector of angle ABC is the only line ℓ such that $C = M_\ell$. This follows easily from Heron's problem.

1.1.25 Draw the line through the incenter I of $\triangle ABC$ and perpendicular to CI. Let this line meet BC and CA at D' and E', respectively (Fig. 87).

Then I is the midpoint of the segment $E'D'$, and it follows from Problem 1.1.10 that $[CDE] \ge [CD'E']$. So, it suffices to show that the area S' of $\triangle CD'E'$ is at

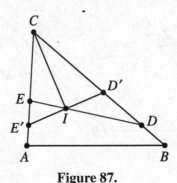

Figure 87.

least $2r^2$. We have $S' = \frac{1}{2}CI \cdot D'E' = CI \cdot D'I$. From the right triangle $D'IC$, we get $CI = r/\sin(C/2)$ and $D'I = r/\cos(C/2)$. Hence

$$S' = \frac{r^2}{\sin\frac{C}{2}\cos\frac{C}{2}} = \frac{2r^2}{\sin C} \geq 2r^2.$$

The equality occurs only if $\angle C = 90°$ and $DE \perp CI$.

1.1.26 It is enough to consider only the points X lying in the half-plane δ determined by the line AB that does not contain the point C. Indeed, if Y is a point in the other half-plane, let X be the reflection of Y in the line AB (Fig. 88). Denote by X_0 the intersection point of the lines CX and AB.

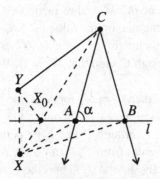

Figure 88.

Then $CX = CX_0 + X_0X = CX_0 + X_0Y \geq CY$. Since $AY = AX$ and $BY = BX$ it follows that $r(X) < r(Y)$. Apart from this, the required point X must lie in the angle ACB. Indeed, if X is situated as in Fig. 88, then $\angle XAB \geq 180° - \angle BAC \geq 90°$, implying $XB > AB$. On the other hand, $XC - XA < AC$,

and we get that

$$r(A) = AB - AC < XB - (XC - XA) = r(X) .$$

Thus, it is enough to consider points X lying in the common part of δ and the angle ACB.

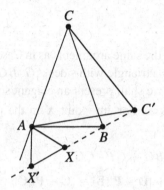

Figure 89.

Let φ be the rotation through $60°$ clockwise about A. Set $C' = \varphi(C)$ and $X' = \varphi(X)$ (Fig. 89). Then $\triangle AXX'$ is equilateral and $AX = XX'$. Moreover, $XC = X'C'$. Hence $r(X) = X'X + BX - C'X'$.

On the other hand, $C'B + BX + XX' \geq C'X'$, which gives $r(X) \geq -C'B$; equality holds precisely when the points X', X, B, and C' lie on a line in this succession. Since $\alpha = \angle BAC \geq 60°$, there are two possible cases.

Case 1. $\alpha = 60°$. Then $C' = B$ and it follows immediately that every point X on the arc $\overset{\frown}{AB}$ of the circumcircle of $\triangle ABC$ gives a solution.

Case 2. $\alpha > 60°$ (Fig. 89). Then $C' \neq B$ and if the points X and X' lie on the line BC', then $\angle AXB = 120°$.

On the other hand, since $\triangle BCC'$ is isosceles and $\angle BCC' = 60° - (180° - 2\alpha) = 2\alpha - 120°$, we have $\angle CBC' = \frac{1}{2}(180° - \angle BCC') = 150° - \alpha$. Hence $\angle ABX = 180° - \angle ABC - \angle CBC' = 30°$ and $\angle BAX = 30°$. This shows that in this case the point X is determined uniquely.

1.1.27 It follows from Pompeiu's theorem (Problem 1.1.6) that the maximum of the distance from O to the third vertex of the equilateral triangle is equal to 2.

1.1.28 Let a, b, c denote the sides of the triangle facing the vertices A, B, C, respectively. We will show that the desired minimum value of the expression

$AP \cdot AG + BP \cdot BG + CP \cdot CG$ is attained when P is the centroid G, and that the minimum value is

$$AG^2 + BG^2 + CG^2$$

$$= \frac{1}{9}\left[(2b^2 + 2c^2 - a^2) + (2c^2 + 2a^2 - b^2) + (2a^2 + 2b^2 - c^2)\right]$$

$$= \frac{1}{3}(a^2 + b^2 + c^2).$$

The problem can be solved using the same arguments as in Case 2 of the solution of Problem 1.1.8. Here $A_0 B_0 C_0$ is the triangle with sides AG, BG, CG, and we leave the details to the reader. Instead, we shall present an elegant solution using the dot product, suggested by M. Klamkin. For any point X in the plane set $\overrightarrow{GX} = \mathbf{X}$. Then

$$AP \cdot AG + BP \cdot BG + CP \cdot CG$$

$$= |\mathbf{A} - \mathbf{P}||\mathbf{A}| + |\mathbf{B} - \mathbf{P}||\mathbf{B}| + |\mathbf{C} - \mathbf{P}||\mathbf{C}|$$

$$\geq |(\mathbf{A} - \mathbf{P}) \cdot \mathbf{A}| + |(\mathbf{B} - \mathbf{P}) \cdot \mathbf{B}| + |(\mathbf{C} - \mathbf{P}) \cdot \mathbf{C}|$$

$$\geq |(\mathbf{A} - \mathbf{P}) \cdot \mathbf{A} + (\mathbf{B} - \mathbf{P}) \cdot \mathbf{B} + (\mathbf{C} - \mathbf{P}) \cdot \mathbf{C}|$$

$$= |\mathbf{A}|^2 + |\mathbf{B}|^2 + |\mathbf{C}|^2 \qquad \text{(since } \mathbf{A} + \mathbf{B} + \mathbf{C} = \mathbf{0})$$

$$= \frac{1}{3}(a^2 + b^2 + c^2),$$

where the last step uses the identity above. Suppose that equality holds. Then

$$|\mathbf{A} - \mathbf{P}||\mathbf{A}| = |(\mathbf{A} - \mathbf{P}) \cdot \mathbf{A}|,$$

$$|\mathbf{B} - \mathbf{P}||\mathbf{B}| = |(\mathbf{B} - \mathbf{P}) \cdot \mathbf{B}|,$$

$$|\mathbf{C} - \mathbf{P}||\mathbf{C}| = |(\mathbf{C} - \mathbf{P}) \cdot \mathbf{C}|.$$

These conditions mean that P lies on each of the lines GA, GB, GC, i.e., $P = G$.

1.1.29 Apply three symmetries with respect to lines to rectangle $ABCD$, as shown in Fig. 90.

Fix an arbitrary point M on the side AB, and consider the point M' on $A''B'''$ such that $A''M' = AM$. Then $MM' = 2AC$. Then show that if N, P, and Q are arbitrary points on the sides BC, CD, and DA, respectively, the perimeter of $MNPQ$ coincides with the length of a broken line connecting M and M'. The latter is minimal when $MN \parallel AC \parallel PQ$ and $NP \parallel BD \parallel QM$. Every parallelogram with these properties inscribed in $ABCD$ (there are infinitely many of them) has a minimal perimeter.

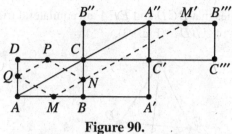

Figure 90.

1.1.30 Let C' and D' be the reflections of C and D in the lines BM and AM, respectively (Fig. 91).

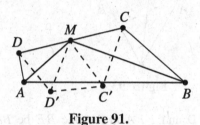

Figure 91.

Then $\triangle C'MD'$ is equilateral, because $C'M = D'M = \frac{1}{2}CD$ and $\angle C'MD' = 180° - 2\angle CMB - 2\angle DMA = 60°$. Hence

$$AD + \frac{1}{2}CD + CB = AD' + D'C' + C'B \geq AB.$$

It follows that $AD + CB \geq AB - \frac{1}{2}CD = 2$. Thus $AB + BC + CD + DA \geq 7$, with equality if and only if C' and D' lie on AB. In the latter case, $\angle ADM = \angle AD'M = 120°$, $\angle BCM = \angle BC'M = 120°$, and $\angle AMD = 60° - \angle CMB = \angle CBM$. Hence triangles AMD and MBC are similar, implying that $AD \cdot BC = (CD/2)^2 = 1$. On the other hand $AD + BC = 2$, and we conclude that $AD = BC = 1$. Therefore the quadrilateral $ABCD$ of minimum perimeter is an isosceles trapezoid with sides $AB = 3$, $BC = AD = 1$, and $CD = 2$ (Fig. 92).

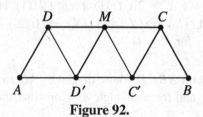

Figure 92.

1.1.31 The hypothesis implies that BCD and EFA are equilateral triangles. Hence BE is an axis of symmetry of $ABDE$ (Fig. 93).

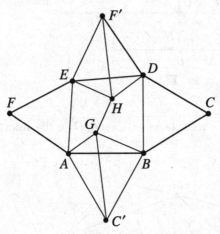

Figure 93.

Let the reflections of BCD and EFA in the line BE be $BC'A$ and $EF'D$, respectively. Since $\angle BGA = 180° - \angle AC'B$, the point G lies on the circumcircle of equilateral triangle ABC'. By Pompeiu's theorem (Problem 1.1.6), $AG + GB = C'G$. Likewise, $DH + HE = HF'$. It follows that

$$CF = C'F' \le C'G + GH + HF' = AG + GB + GH + DH + HE,$$

with equality if and only if G and H both lie on $C'F'$.

1.1.32

(a) Let $ABCD$ be a convex quadrilateral. It follows from the triangle inequality that $XA + XB + XC + XD \ge AC + BD$, with equality only when X coincides with the intersection point of the diagonals AC and BD.

(b) Let O be the center of symmetry of the given polygon $A_1 A_2 \ldots A_n$. For any point X in the plane, let X' be the reflection of X in O. Then for any $i = 1, 2, \ldots, n$ we have $A_i X + A_i X' \ge 2 A_i O$ (Problem 1.1.11). Hence for $t(X) = \sum_{i=1}^{n} A_i X$, it follows that $t(X) = t(X')$ and $t(X) = \frac{1}{2}(t(X) + t(X')) \ge t(O)$, where equality holds only for $X = O$.

1.1.33 Consider any quadrilateral $ABCD$ whose diagonals AC and BD have given lengths a and b, respectively, and form an angle α. Construct the parallelograms $ABDM$ and $BCKD$ (Fig. 94).

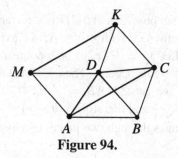

Figure 94.

Then the quadrilateral $ACKM$ is a parallelogram. Indeed, $AM \parallel BD$ and $BD \parallel CK$ imply $AM \parallel CK$; in addition, $AM = BD = CK$. This parallelogram is completely determined by its sides $AC = MK = a$, $AM = CK = b$, and $\angle CAM = \alpha$.

Note now that since $DM = AB$ and $DK = BC$, the perimeter of $ABCD$ is equal to the sum $DA + DC + DK + DM$, that is, to the sum of distances from the point D to the vertices of the parallelogram $ACKM$.

It follows from Problem 1.1.32 (a) that the perimeter of $ABCD$ is minimal when D is the intersection point of the diagonals AK and CM of $ACKM$. Tracing backward the construction from above, we conclude that in the latter case the original quadrilateral $ABCD$ is a parallelogram with diagonals of lengths a and b forming the given angle α.

1.1.34 Note first that $[ABM] + [CDM] = \frac{1}{2}[ABCD] = \frac{1}{2}S$. Construct the point Q outside $ABCD$ such that $AQ = CM$ and $BQ = DM$ (Fig. 95).

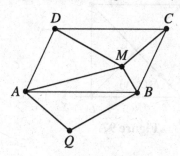

Figure 95.

Then $\triangle ABQ \cong \triangle CDM$, so $[AQBM] = [ABM] + [CDM] = \frac{1}{2}S$. On the other hand, $[AQBM] = [AMQ] + [BMQ]$. Since $AM \cdot AQ \geq 2[AMQ]$ and $BM \cdot BQ \geq 2[BMQ]$, we obtain

$$AM \cdot CM + BM \cdot DM = AM \cdot AQ + BM \cdot BQ$$
$$\geq 2([AMQ] + [BMQ]) = 2[AQBM] = S.$$

To deal with the equality case, suppose that $ABCD$ is a rectangle with $AB = a$, $BC = b$. Set a coordinate system Axy with origin A, Ax axis the ray AB, and Ay axis the ray AD (Fig. 96). Let $M = M(x, y)$ be a point for which the equality $AM \cdot CM + BM \cdot DM = S = ab$ occurs. Then $\angle MAQ = \angle MBQ = 90°$ and since $\angle BAQ = \angle MCQ$, we get $\angle MAB = \angle MCB$. On the other hand, $\tan \angle MAB = \frac{y}{x}$, $\tan \angle MCB = \frac{a-x}{b-y}$ and therefore $y(b - y) = x(a - x)$.

If $a \neq b$ then the point M ranges through two pieces of a hyperbola (see Fig. 96 for the case $a > b$).

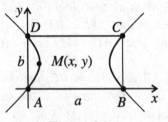

Figure 96.

If $a = b$, i.e., $ABCD$ is a square, then $y(a - y) = x(a - x)$, which gives $x = y$ or $x + y = a$. Hence in this case M ranges over the diagonals AC and BD of this square (Fig. 97).

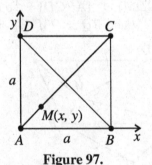

Figure 97.

1.1.35 If $ABCD$ is the given quadrilateral, consider the quadrilateral AB_1CD, where B_1 is the reflection of B in the perpendicular bisector of diagonal AC. Clearly, $ABCD$ and AB_1CD have the same areas, and the sides of AB_1CD are b, a, c, d, in this order. Hence $S = [B_1CD] + [DAB_1] \leq \frac{1}{2}(ac + bd)$. Equality occurs if and only if $\angle DAB_1 = \angle B_1CD = 90°$. This condition means that AB_1CD is a cyclic quadrilateral with two opposite right angles. Equivalently, $ABCD$ is also cyclic (having the same circumcircle), and its diagonals are perpendicular.

1.1.36 Let M and N be the midpoints of AB and CD. Since $\triangle ABD \cong \triangle BAC$ we get $MD = MC$, which implies $MN \perp CD$ (Fig. 98). In a similar way one gets $MN \perp AB$.

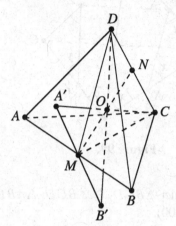

Figure 98.

Let φ be the rotation in space through $180°$ about the line MN. Then $\varphi(A) = B$, $\varphi(B) = A$, $\varphi(C) = D$, and $\varphi(D) = C$.

Let X be an arbitrary point in space that is not on MN. Set $X' = \varphi(X)$ and let Y be the midpoint of the segment XX'. Then Y lies on the line MN and $t(X') = t(X)$. Since $PX + PX' > 2PY$ for any point P (Problem 1.1.11), we have $2t(X) = t(X) + t(X') > 2t(Y)$. This shows that it is enough to consider only points X on the line MN.

Let ψ be the rotation about the line MN that maps A and B to points A' and B' on the plane CDM such that the quadrilateral $A'B'CD$ is convex. Then for any X on the line MN we have $t(X) = A'X + B'X + CX + DX$, and Problem 1.1.32 (a) implies that $t(X)$ is a minimum when X coincides with the intersection point O of the diagonals $A'C$ and $B'D$. The point O is characterized by the condition $\angle AOB = \angle COD$.

1.1.37 *Hint.* Let B' be the reflection of B in the plane α. Show that the required line is the intersection line of α and the plane OAB'.

1.1.38

(a) The statement follows easily from Problem 1.1.2 (Fig. 99).

(b) Let α, β, γ, and δ be the sums of the face angles of the tetrahedron $ABCD$ at the vertices A, B, C, and D, respectively. Then the given condition implies $\alpha + \gamma = 360° = \beta + \delta$. We may assume that $\alpha \leq 180°$ and $\beta \leq 180°$. In

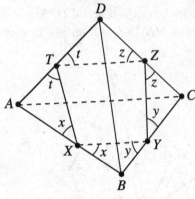

Figure 99.

the plane of $\triangle ABC$ construct $\triangle BCD' \cong \triangle BCD$, $\triangle ABD'' \cong \triangle ABD$, and $\triangle AD''C' \cong \triangle ADC$ (Fig. 100).

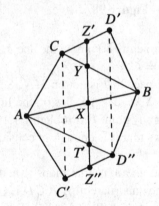

Figure 100.

It is now easy to see that the quadrilateral $C'D''D'C$ is a parallelogram lying entirely in the hexagon $AC'D''BD'C$. Moreover, it follows from $\triangle C'CA$ that $CC' = 2AC \sin \frac{\alpha}{2}$. For any point X on AB lying in the parallelogram $C'D''D'C$, the line through X that is parallel to CC' intersects the lines BC, CD', AD'', and $C'D''$ at points Y, Z', T', and Z'', respectively. Now construct points Z on CD and T on AD such that $CZ = CZ'$ and $AT = AT'$. Then the length of the broken line $XYZTX$ is $Z'Z'' = CC'$. We leave it to the reader to show that the length of any such broken line is not less than CC' and is equal to CC' when X lies in the parallelogram $C'D''D'C$ and the points Y, Z, and T are obtained from X in the way described above.

1.1.39 Let A and B be the two cities and let ℓ_1 and ℓ_2 be the (parallel) banks of the river, where ℓ_1 is between A and ℓ_2 (Fig. 101).

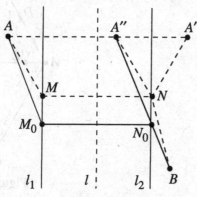

Figure 101.

Construct the line $\ell \parallel \ell_1$ such that ℓ_1 and ℓ_2 are symmetric with respect to ℓ, the reflection A' of A in ℓ, and the reflection A'' of A' in ℓ_2. Next, let N_0 be the intersection point of ℓ_2 and BA'', and let M_0 be the point on ℓ_1 such that $M_0N_0 \perp \ell_1$. Let $M \in \ell_1$ and $N \in \ell_2$ be arbitrary points such that $MN \perp \ell_1$. Then $AM = A'N = A''N$ and therefore

$$AM + MN + NB = A''N + NB + M_0N_0 \geq A''B + M_0N_0,$$

where equality holds when $N = N_0$. Clearly the latter implies $M = M_0$. Thus, the road AM_0N_0B has the shortest possible length.

1.1.40

(a) First, we will show that the best strategy for James is to choose $Y = B$ or $Y = C$. For any point X on AC consider its reflection X' in AB and the reflection X'' of X' in BC (Fig. 102).

Clearly (see Problem 1.1.1), if X and Y are already chosen, John has to choose Z as the intersection point of AB and $X'Y$. For such a choice of Z we have

$$XY + YZ + ZX = XY + YX' = XY + YX''.$$

Since Y lies on BC, the latter sum will be a maximum when $Y = B$ or $Y = C$, depending on the position of the segment XX''.

Next, if James chooses $Y = B$, then John will choose $Z = B$, and the perimeter of $\triangle XYZ$ will be $2XB$. In case James takes $Y = C$, John will put Z at the intersection point of $X'C$ and AB (Fig. 103), and then the perimeter of $\triangle XYZ$

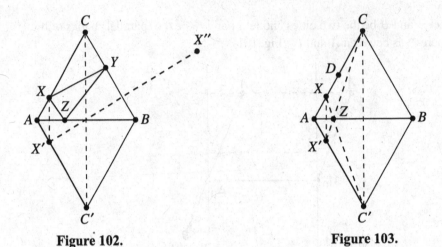

Figure 102.   **Figure 103.**

will be $XC + X'C = XC + XC'$. Let D be the midpoint of AC. Clearly John has to choose X on the segment DC. We leave to the reader to show that there exists a point E on DC such that $2BE = CE + C'E$ and that John has to choose $X = E$.

(b) For any choice of X on AC, James has to choose Y on BC such that $XY \parallel AB$ (Fig. 104).

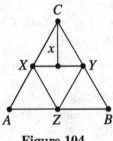

Figure 104.

Then for any Z on AB we have $[XYZ] = \frac{1}{2} XY(h - x) = \frac{x(h-x)}{2}$, where $h = \frac{\sqrt{3}}{2}$ is the length of the altitude in $\triangle ABC$ and x is the distance from C to XY. The quadratic function $x(h - x)$ of x has a maximum at $x = \frac{h}{2}$, i.e., when X is the midpoint of AC. Then $[XYZ] = \frac{\sqrt{3}}{16}$. This is the maximum area that John can achieve, and his strategy is to put X at the midpoint of AC.

1.1.41 Through A_0, B_0, C_0, draw lines parallel to B_1C_1, C_1A_1, A_1B_1, respectively. These form the sides BC, CA, AB of a $\triangle ABC$ similar to $\triangle A_1B_1C_1$. Now suppose each of the lines drawn is rotated about A_0, B_0, C_0, respectively, by the same amount. Then they meet at the same angles as before, always forming triangles similar to $\triangle A_1 B_1 C_1$. The triangle of maximum area among them is the one whose sides have maximal length.

To find it, recall that the locus of points B such that $\angle A_0 B C_0$ has a given measure β is an arc of a circle with chord $A_0 C_0$. This suggests that we construct the circumcircles of $\triangle A_0 C_0 B$, $\triangle B_0 A_0 C$, and $\triangle B_0 C_0 A$ (Fig. 105).

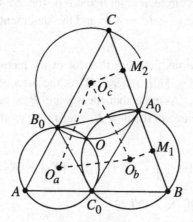

Figure 105.

Denote their centers by O_b, O_c, and O_a, respectively. It is easy to prove that these circumcircles have a point O in common.

We show next that $\triangle O_a O_b O_c \sim \triangle ABC$. Indeed,

$$\angle C = \frac{1}{2} \overset{\frown}{A_0 O B_0} \qquad \text{and} \qquad \angle O_a O_c O_b = \frac{1}{2} \overset{\frown}{A_0 O} + \frac{1}{2} \overset{\frown}{O B_0},$$

because $O_c O_a$ and $O_c O_b$ bisect arcs $\overset{\frown}{B_0 O}$ and $\overset{\frown}{A_0 O}$, respectively. So $\angle C = \angle O_a O_c O_b$. Similarly, $\angle A = \angle O_c O_a O_b$, $\angle B = \angle O_a O_b O_c$. Therefore $\triangle O_a O_b O_c \sim \triangle ABC \sim \triangle A_1 B_1 C_1$.

Finally, we show that the largest triangle ABC through the points A_0, B_0, C_0 is the one whose sides are parallel to those of triangle $O_a O_b O_c$.

To prove this, note that the perpendiculars from O_b and O_c bisect the chords BA_0 and CA_0 at M_1 and M_2, and so $M_1 M_2 = \frac{1}{2} BC$. The line segment $M_1 M_2$ is the orthogonal projection of $O_b O_c$ on BC and is largest when $BC \parallel O_b O_c$. Since $\triangle O_a O_b O_c \sim \triangle ABC$, all three sides of the maximal triangle are parallel to those of $\triangle O_a O_b O_c$.

Thus, to construct the maximal triangle, first construct any triangle through A_0, B_0, C_0 similar to $\triangle A_1 B_1 C_1$. Then construct the centers O_a, O_b, O_c of the circumcircles of $\triangle A_0 C_0 B$, $\triangle B_0 A_0 C$, and $\triangle B_0 C_0 A$. Finally, construct lines through A_0, B_0, C_0 parallel to $O_b O_c$, $O_c O_a$, $O_a O_b$, respectively. They form the sides BC, CA, AB of the desired maximal triangle.

4.2   Employing Algebraic Inequalities

1.2.6 Let R be the radius of the circle, and let a and b be the lengths of the sides of a rectangle inscribed in it. Then $a^2 + b^2 = 4R^2$ and the statement follows from the inequality $ab \le \frac{a^2+b^2}{2}$.

1.2.7 It follows from the previous problem that for every rectangle Π inscribed in a circle K we have $\frac{\pi}{2}[\Pi] \le [K]$, where $[\Pi]$ is the area of Π and $[K]$ that of the disk determined by K. Assume that the square P is cut into rectangles $\Pi_1, \Pi_2, \ldots, \Pi_n$. Then for the area of the disk K determined by the circumcircle of P we have

$$[K] = \frac{\pi}{2}[P] = \frac{\pi}{2}([\Pi_1] + [\Pi_2] + \cdots + [\Pi_n]) \le [K_1] + [K_2] + \cdots + [K_n],$$

where K_i is the circumcircle of Π_i, $1 \le i \le n$.

1.2.8 Let a, b, and c be the lengths of the edges of a rectangular parallelepiped with a given volume V. Then $abc = V$, and the arithmetic mean–geometric mean inequality gives

$$\frac{S}{6} = \frac{ab + bc + ca}{3} \ge \sqrt[3]{(abc)^2} = V^{2/3},$$

where S is the surface area of the parallelepiped. This shows that the minimum of S is attained when $a = b = c$.

1.2.9 Let S_1, S_2, S_3, and S_4 be the areas of the rectangles, where $S_1, S_2, S_3 \ge 1$ and $S_4 \ge 2$ (Fig. 106). Then $S_1 S_4 = S_2 S_3$ and $S_2 + S_3 \ge 2\sqrt{S_2 S_3} = 2\sqrt{S_1 S_4} \ge 2\sqrt{2}$.

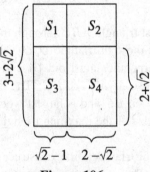

Figure 106.

Hence $S_1 + S_2 + S_3 + S_4 \geq 3 + 2\sqrt{2}$, i.e., $d \geq 3 + 2\sqrt{2}$. It is shown in Fig. 106 how to cut a rectangle with side lengths 1 and $3 + 2\sqrt{2}$ in the required way.

1.2.10 Let a and h_a be the lengths of a side and the corresponding altitude in the given triangle. Then its perimeter is larger than $a + 2h_a$. Let c be the length of the side of the square. Then $\frac{ah_a}{2} = c^2$ and $a + 2h_a \geq 2\sqrt{2ah_a} = 4c$. So, the perimeter of the triangle is larger than the perimeter of the square.

1.2.11

(a) First we will find the shortest segment that cuts off a triangle of area S from an angle of measure α. Consider an arbitrary line that cuts off a triangle of area S from the given angle.

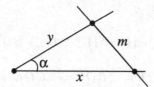

Figure 107.

Let x and y be the lengths of the segments cut off from the sides of the angle and let m be the length of the third side of the triangle obtained (Fig. 107). The law of cosines gives $m^2 = x^2 + y^2 - 2xy \cos \alpha$. Using the inequality $x^2 + y^2 \geq 2xy$, it follows that $m^2 \geq 2xy(1 - \cos \alpha)$. Since $2S = xy \sin \alpha$, one gets $m^2 \geq 4S \frac{(1 - \cos \alpha)}{\sin \alpha} = 4S \tan \frac{\alpha}{2}$. Hence the shortest segment with the required property has length $m = \sqrt{4S \tan \frac{\alpha}{2}}$.

One concludes that the solution of the problem is given by a segment of length $2\sqrt{S \tan \frac{\alpha}{2}}$, where S is the area of the triangle and α is the measure of its smallest angle.

(b) Let s be the semiperimeter of the given triangle. Using the same notation as in (a) we have $x + y = s$ and $m^2 = x^2 + y^2 - 2xy \cos \alpha$. Then

$$m^2 = (x + y)^2 - 2xy(1 + \cos \alpha) \geq (x + y)^2 - \frac{(x + y)^2}{2}(1 + \cos \alpha)$$

$$= \frac{(x + y)^2(1 - \cos \alpha)}{2} = \left((x + y) \sin \frac{\alpha}{2}\right)^2 = \left(s \sin \frac{\alpha}{2}\right)^2,$$

i.e., $m \geq s \sin \frac{\alpha}{2}$. Hence in this case the solution of the problem is given by a segment of length $s \sin \frac{\alpha}{2}$, where s is the semiperimeter of the triangle and α is the smallest angle.

1.2.12 We have
$$2[AOB] \le AO \cdot BO \le \frac{AO^2 + BO^2}{2},$$
with equality if and only if $\angle AOB = 90°$ and $AO = BO$. Likewise,
$$2[BOC] \le \frac{BO^2 + CO^2}{2},$$
$$2[COD] \le \frac{CO^2 + DO^2}{2},$$
$$2[DOA] \le \frac{DO^2 + AO^2}{2}.$$

Adding up these inequalities yields

$$2\Big([AOB] + [BOC] + [COD] + [DOA]\Big) \le AO^2 + BO^2 + CO^2 + DO^2,$$

with equality if and only if $\angle AOB = \angle BOC = \angle COD = \angle DOA = 90°$ and $AO = BO = CO = DO$. On the other hand, for any quadrilateral $ABCD$ (convex or not) and any point O we have

$$2[ABCD] \le 2\Big([AOB] + [BOC] + [COD] + [DOA]\Big).$$

It readily follows that $ABCD$ is a square with center O.

1.2.13

(a) Let $ABCD$ be a convex quadrilateral of area 1. Then

$$1 = [ABD] + [BCD] \le \frac{1}{2}(AB \cdot AD + BC \cdot CD),$$

$$1 = [ABC] + [ACD] \le \frac{1}{2}(AB \cdot BC + AD \cdot CD).$$

Adding up gives
$$(AB + CD)(AD + BC) \ge 4,$$

and now the arithmetic mean–geometric mean inequality implies

$$AB + CD + AD + BC \ge 2\sqrt{(AB + CD)(AD + BC)} \ge 4.$$

Hence the minimum of the perimeter of $ABCD$ is 4, and it is attained only if $ABCD$ is a square.

(b) The area of $ABCD$ is given by

$$1 = [ABCD] = \frac{1}{2}AC \cdot BD \sin \varphi,$$

where φ is the angle between the diagonals AC and BD. Hence $AC \cdot BD \geq 2$, and it follows from the arithmetic mean–geometric mean inequality that

$$AC + BD \geq 2\sqrt{AC \cdot BD} \geq 2\sqrt{2},$$

with equality only if $AC \perp BD$ and $AC = BD$.

1.2.14 Let $ABCD$ be a quadrilateral with area 32 and $AB + BD + DC = 16$. Its area can be expressed as

$$[ABCD] = \frac{1}{2}AB \cdot BD \sin \angle ABD + \frac{1}{2}DC \cdot BD \sin \angle CDB.$$

Using the fact that the sine of an angle does not exceed 1, and also the arithmetic mean–geometric mean inequality, we obtain

$$32 = [ABCD] \leq \frac{1}{2}AB \cdot BD + \frac{1}{2}DC \cdot BD = \frac{1}{2}BD(AB + CD)$$

$$\leq \frac{1}{2}\left(\frac{BD + AB + CD}{2}\right)^2 = 32.$$

Therefore the conditions of the problem statement are met only if all inequalities above are equalities, that is,

$$\angle ABD = \angle CDB = 90° \quad \text{and} \quad BD = AB + CD = 8.$$

It is straightforward (Fig. 108) that in the latter case there is only one possible value for the diagonal AC, namely

$$AC = \sqrt{BD^2 + (AB + CD)^2} = 8\sqrt{2}.$$

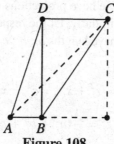

Figure 108.

1.2.15 Let a, b, and c be the lengths of the edges of the right trihedral angle of a terahedron. The sum of its six edges is

$$s = a + b + c + \sqrt{a^2 + b^2} + \sqrt{b^2 + c^2} + \sqrt{c^2 + a^2}.$$

It follows from the inequality $\sqrt{\frac{x^2+y^2}{2}} \geq \frac{x+y}{2}$ that $s \geq (1 + \sqrt{2})(a + b + c)$. Now the arithmetic mean–geometric mean inequality gives

$$s \geq 3(1 + \sqrt{2})\sqrt[3]{abc} = 3(1 + \sqrt{2})\sqrt[3]{6V},$$

where V is the volume of the tetrahedron. Thus, the required tetrahedron is the one with $a = b = c = \frac{s}{3(1+\sqrt{2})}$.

1.2.16 Suppose first that the parallelepiped is rectangular with edge lengths x, y, z. Then, by the arithmetic mean–geometric mean inequality,

$$2(xy + yz + zx) \geq 6(216)^{2/3} = 216.$$

So the surface area is at least 216, with equality if and only if $x = y = z$, i.e., the parallelepiped is a cube. Now consider a nonrectangular parallelepiped whose "top" face is not directly above its "bottom" face. Then moving the top face above the bottom one leaves the volume fixed and decreases the surface area. Repeating this for each pair of opposite faces yields a rectangular parallelepiped with strictly smaller surface area and the same volume 216. By the previous part, this rectangular parallelepiped has surface area at least 216, so the original parallelepiped has surface area greater than 216. Thus if a parallelepiped has volume 216 and surface area 216, it must be a cube.

1.2.17 Let M be the intersection point of the segment AB with the plane α and let $a = AM, b = BM$. Consider a sphere through A and B and denote by x and y the lengths of the parts into which M divides the diameter of the disk that the sphere cuts off from α. Then $xy = ab$ and $x + y \geq 2\sqrt{xy} = 2\sqrt{ab}$. Thus the disk is of minimum area only if $x = y = \sqrt{ab}$.

1.2.18 Let the ith side of the broken line have projections of lengths x_i, y_i, z_i onto the axes Ox, Oy, Oz, respectively. Similarly, let the respective projections of this side onto the planes Oyz, Ozx, Oxy have lengths a_i, b_i, c_i. Denote by l_i the length of the ith side itself. Then

$$a_i^2 = y_i^2 + z_i^2, \quad b_i^2 = z_i^2 + x_i^2, \quad c_i^2 = x_i^2 + y_i^2,$$

$$l_i^2 = x_i^2 + y_i^2 + z_i^2 = \frac{1}{2}(a_i^2 + b_i^2 + c_i^2).$$

Then, by the arithmetic mean–quadratic mean inequality,

$$a_i + b_i + c_i \leq 3\sqrt{\frac{a_i^2 + b_i^2 + c_i^2}{3}} = 3\sqrt{\frac{2l_i^2}{3}} = l_i\sqrt{6}.$$

Adding up all such inequalities gives $a + b + c \leq l\sqrt{6}$. This inequality becomes equality for, say, the line segment (which is an open broken line) with endpoints $(0, 0, 0)$ and $(1, 1, 1)$.

(b) There exists a closed broken line with the given property. An example is the line joining the points

$$(0, 0, 0), \ (1, 1, 1), \ (2, 2, 0), \ (3, 1, -1), \ (2, 0, -2), \ (1, -1, -1), \ (0, 0, 0)$$

in this order.

1.2.19 We have $ax + by + cz = 2[ABC]$. Then the Cauchy–Schwarz inequality implies that

$$(ax + by + cz)\left(\frac{a}{x} + \frac{b}{y} + \frac{c}{z}\right) \geq (a + b + c)^2.$$

Hence

$$\frac{a}{x} + \frac{b}{y} + \frac{c}{z} \geq \frac{(a + b + c)^2}{2[ABC]}$$

with equality only if $x = y = z$. Thus the desired point X is the incenter of $\triangle ABC$.

(b) The Cauchy–Schwarz inequality gives

$$(ax + by + cz)\left(\frac{1}{ax} + \frac{1}{by} + \frac{1}{cz}\right) \geq 9.$$

Hence

$$\frac{1}{ax} + \frac{1}{by} + \frac{1}{cz} \geq \frac{9}{2[ABC]}$$

with equality only if $ax = by = cz$. Show that the only point X with this property is the centroid of $\triangle ABC$.

1.2.20 Denote by h_1, h_2, h_3, h_4 the lengths of the altitudes of the tetrahedron $ABCD$. Then

$$\frac{d_1}{h_1} + \frac{d_2}{h_2} + \frac{d_3}{h_3} + \frac{d_4}{h_4} = \frac{\text{Vol}(ABCX)}{\text{Vol}(ABCD)} + \cdots + \frac{\text{Vol}(DABX)}{\text{Vol}(ABCD}} = 1$$

and the arithmetic mean–geometric mean inequality gives

$$1 = \frac{d_1}{h_1} + \frac{d_2}{h_2} + \frac{d_3}{h_3} + \frac{d_4}{h_4} \geq 4\sqrt[4]{\frac{d_1 d_2 d_3 d_4}{h_1 h_2 h_3 h_4}}.$$

Hence $d_1 d_2 d_3 d_4 \leq \frac{h_1 h_2 h_3 h_4}{256}$, where equality occurs only if $d_i = \frac{h_i}{4}$, $1 \leq i \leq 4$. This shows that the product $d_1 d_2 d_3 d_4$ is a maximum only if X is the centroid of $ABCD$.

1.2.21 Using the fact that the triangles under consideration are similar to triangle ABC, one easily obtains that

$$[ABC] = (\sqrt{S_1} + \sqrt{S_2} + \sqrt{S_3})^2.$$

Now the arithmetic mean–geometric mean inequality gives

$$S_1 + S_2 + S_3 \geq \frac{1}{3}(\sqrt{S_1} + \sqrt{S_2} + \sqrt{S_3})^2 = \frac{[ABC]}{3},$$

where equality is attained only if $S_1 = S_2 = S_3 = \frac{[ABC]}{9}$. This implies easily that the sum $S_1 + S_2 + S_3$ is a minimum only if X is the centroid of ABC.

1.2.22 Set $[A_1 A_2 M] = S_1, [B_1 B_2 M] = S_2, [C_1 C_2 M] = S_3, [A_1 C_2 M] = T_1, [B_1 A_2 M] =$? $[C_1 B_2 M] = T_3$ (Fig. 109).

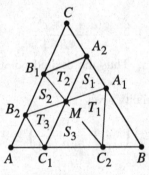

Figure 109.

Then $S_1 S_2 S_3 = T_1 T_2 T_3$. The arithmetic mean–geometric mean inequality, used twice, gives

$$\frac{1}{S_1} + \frac{1}{S_2} + \frac{1}{S_3} \geq \frac{3}{\sqrt[3]{S_1 S_2 S_3}} = \frac{3}{\sqrt[6]{S_1 S_2 S_3 T_1 T_2 T_3}}$$

$$\geq \frac{18}{S_1 + S_2 + S_3 + T_1 + T_2 + T_3} \geq \frac{18}{[ABC]}.$$

Hence the least value of the given sum is equal to $18/[ABC]$. This minimum value is attained only if $S_1 = S_2 = S_3 = T_1 = T_2 = T_3 = [ABC]/6$, i.e., if M is the centroid of $\triangle ABC$ and the three lines contain the medians of the triangle (Fig. 110).

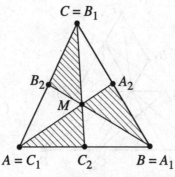

Figure 110.

1.2.23 Set $\lambda = \frac{AC_1}{C_1B}$, $\mu = \frac{BA_1}{A_1C}$, and $\nu = \frac{CB_1}{B_1A}$. According to Ceva's theorem (cf. Glossary), $\lambda\mu\nu = 1$. On the other hand,

$$\frac{[AB_1C_1]}{[ABC]} = \frac{AC_1}{AB} \cdot \frac{AB_1}{AC} = \frac{\lambda}{(\lambda+1)(\nu+1)},$$

$$\frac{[BA_1C_1]}{[ABC]} = \frac{BA_1}{BC} \cdot \frac{BC_1}{BA} = \frac{\mu}{(\mu+1)(\lambda+1)},$$

$$\frac{[CB_1A_1]}{[ABC]} = \frac{CB_1}{CA} \cdot \frac{CA_1}{CB} = \frac{\nu}{(\nu+1)(\mu+1)}.$$

Hence

$$\frac{[A_1B_1C_1]}{[ABC]} = 1 - \frac{\lambda}{(\lambda+1)(\mu+1)} - \frac{\mu}{(\mu+1)(\lambda+1)} - \frac{\nu}{(\nu+1)(\mu+1)}$$

$$= \frac{1+\lambda\mu\nu}{(\lambda+1)(\mu+1)(\nu+1)} = \frac{2}{(\lambda+1)(\mu+1)(\nu+1)}.$$

Multiplying the inequalities $1 + \lambda \geq 2\sqrt{\lambda}$, $1 + \mu \geq 2\sqrt{\mu}$, and $1 + \nu \geq 2\sqrt{\nu}$ gives $(1+\lambda)(1+\mu)(1+\nu) \geq 8$. Thus $[A_1B_1C_1] \leq \frac{1}{4}[ABC]$, where equality holds when $\lambda = \mu = \nu = 1$, i.e., when X is the centroid of the triangle.

Hence the area of $\triangle A_1B_1C_1$ is a maximum if X is the centroid of $\triangle ABC$.

1.2.24 Draw the lines through P and parallel to the sides of $\triangle ABC$ as shown in Fig. 111.

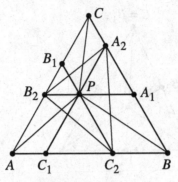

Figure 111.

Then $PA = B_2C_2$, $PB = C_2A_2$, and $PC = A_2B_2$. Hence we have to prove that $[A_2B_2C_2] \leq \dfrac{1}{3}[ABC]$. To do this note that

$$[A_2B_2C_2] = [A_2B_2P] + [B_2C_2P] + [C_2A_2P]$$

$$= \frac{1}{2}([A_2CB_1P] + [B_2AC_1P] + [C_2BA_1P])$$

$$= \frac{1}{2}([ABC] - [A_1A_2P] - [B_1B_2P] - [C_1C_2P]).$$

Hence Problem 1.2.21 implies that $[A_2B_2C_2] \leq \frac{1}{3}[ABC]$.

1.2.25 Denote the inradii in question by r_a, r_b, r_c. Then

$$r_a = \frac{2[AB_1C_1]}{AB_1 + AC_1 + B_1C_1} = \frac{\sqrt{3}}{2} \cdot \frac{AB_1 \cdot AC_1}{AB_1 + AC_1 + B_1C_1}.$$

The law of cosines for $\triangle AB_1C_1$ gives

$$B_1C_1 = \sqrt{AB_1^2 + AC_1^2 - AB_1 \cdot AC_1} \geq \sqrt{AB_1 \cdot AC_1}.$$

Then

$$r_a \leq \frac{\sqrt{3}}{2} \cdot \frac{AB_1 \cdot AC_1}{2\sqrt{AB_1 \cdot AC_1} + \sqrt{AB_1 \cdot AC_1}}$$

$$= \frac{\sqrt{3}}{6}\sqrt{AB_1 \cdot AC_1} \leq \frac{1}{4\sqrt{3}} \cdot (AB_1 + AC_1).$$

By symmetry, analogous inequalities hold true for r_b and r_c. Now adding up leads to

$$r_a + r_b + r_c \le \frac{1}{4\sqrt{3}}(AB_1 + AC_1 + BC_1 + BA_1 + CA_1 + CB_1)$$

$$= \frac{1}{4\sqrt{3}}(AB + BC + CA).$$

Clearly, equality occurs only if A_1, B_1, C_1 are the midpoints of the respective sides.

1.2.26 For brevity, let $[DBK] = [KBM] = [MBE] = S$ (Fig. 112). Then

$$\frac{[ABT]}{S} = \frac{AB \cdot BT}{DB \cdot BK}, \quad \frac{[TBP]}{S} = \frac{TB \cdot BP}{KB \cdot BM}, \quad \frac{[PBC]}{S} = \frac{PB \cdot BC}{MB \cdot BE}.$$

It follows by the arithmetic mean–geometric mean inequality that

$$\frac{[ABC]}{S} = \frac{[ABT] + [TBP] + [PBC]}{S}$$

$$\ge 3\sqrt[3]{\frac{AB \cdot BT}{DB \cdot BK} \cdot \frac{TB \cdot BP}{KB \cdot BM} \cdot \frac{PB \cdot BC}{MB \cdot BE}}$$

$$= 3\left(\frac{TB \cdot BP}{KB \cdot BM}\right)^{2/3} \left(\frac{AB \cdot BC}{DB \cdot BE}\right)^{1/3}$$

$$= 3\left(\frac{[TBP]}{S}\right)^{2/3} \left(\frac{[ABC]}{[DBE]}\right)^{1/3}.$$

Since $[DBE] = 3S$, the inequality obtained above can be rewritten as $[ABC] \ge 3[TBP]$, implying the desired $AC \ge 3TP$.

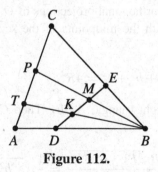

Figure 112.

1.2.27 Assume without loss of generality that $A \ge B \ge C$. Then $a \ge b \ge c$ and the Chebyshev inequality (cf. Glossary) gives

$$\Delta = \frac{aA + bB + cC}{a + b + c} \ge \frac{(a+b+c)(A+B+C)}{3(a+b+c)} = \frac{1}{3}(A+B+C) = \frac{\pi}{3}.$$

Hence the minimum of Δ is $\pi/3$, and it is attained only if the triangle is equilateral.

We shall show that Δ does not have a maximum if only nondegenerate triangles are considered. To make sure, note first that the triangle inequality gives

$$a + b + c > 2a, a + b + c > 2b, a + b + c > 2c.$$

This implies

$$\Delta = \frac{aA + bB + cC}{a + b + c} < \frac{A + B + C}{2} = \frac{\pi}{2}.$$

We now show that $\pi/2$ is a sharp upper bound for Δ. Consider an isosceles triangle ABC such that $AC = BC = 1$ and $\angle BAC = \angle ABC = x$, where $0 < x < \pi/2$. Then $AB = 2\cos x$ and we get

$$\Delta(x) = \frac{x + (\pi - 2x)\cos x}{1 + \cos x}.$$

Hence $\Delta(x)$ can be made arbitrarily close to $\pi/2$, since $\lim_{x \to 0} \Delta(x) = \pi/2$.

1.2.28 Let R be the radius of the sphere and h the length of the altitude of the cone. Then the volume V of the cone is given by $V = \frac{\pi h^2(2R-h)}{3}$. By the arithmetic mean–geometric mean inequality,

$$V = \frac{4\pi}{3} \cdot \frac{h}{2} \cdot \frac{h}{2}(2R - h) \le \frac{4\pi}{3}\left(\frac{2R}{3}\right)^3,$$

with equality only if $h/2 = 2R - h$, i.e., $h = \frac{4R}{3}$.

1.2.29 Let O be the center of the given sphere and R its radius. Set $PA = a$, $PB = b$, and $PC = c$. Since the orthogonal projections of O on the plane (PAB) and on the line PC coincide with the midpoints of the segments AB and PC respectively, we get that

$$(1) \qquad\qquad a^2 + b^2 + c^2 = 4R^2.$$

Let PH be the altitude of the right triangle APB. Then CH is the altitude of triangle ACB. Hence

$$(2) \qquad [ABC] = \frac{AB.CH}{2} = \frac{AB\sqrt{PC^2 + PH^2}}{2} = \frac{1}{2}\sqrt{a^2b^2 + b^2c^2 + c^2a^2}.$$

Now the inequality $3(a^2b^2 + b^2c^2 + c^2a^2) \le (a^2 + b^2 + c^2)^2$ together with (1) and (2) implies that $[ABC] \le \frac{2R^2}{\sqrt{3}}$, i.e., the maximum area of triangle ABC is equal to $\frac{2R^2}{\sqrt{3}}$. It is attained if and only if $PA = PB = PC = \frac{2R}{\sqrt{3}}$.

1.2.30 It is easy to prove that the ratio of the volumes of two tetrahedra with a common trihedral angle is equal to the ratio of the products of the lengths of their edges forming this trihedral angle. Hence the tetrahedron $OABC$ has a maximum volume when the product $OA \cdot OB \cdot OC$ is a maximum. Since $OA + OB + OC = a$, it follows from the arithmetic mean–geometric mean inequality that $OA \cdot OB \cdot OC \leq \left(\frac{a}{3}\right)^3$, where equality holds when $OA = OB = OC = \frac{a}{3}$. This is the case when the volume of the tetrahedron is a maximum.

1.2.31 Fix a point M_0 on the face ABC and let A_0, B_0, and C_0 be the feet of the perpendiculars from M_0 to the planes BCD, ACD, and ABD, respectively. Let V_0 and V be the volumes of the tetrahedra $M_0A_0B_0C_0$ and $MA_1B_1C_1$, and let $x = MA_1$, $y = MB_1$, and $z = MC_1$. Since the trihedral angles at the vertices M_0 and M of these tetrahedra are congruent, it follows that

$$\frac{V}{V_0} = \frac{xyz}{M_0A_0 \cdot M_0B_0 \cdot M_0C_0}.$$

Thus, M must be chosen such that xyz is a maximum. Let S_A, S_B, and S_C be the areas of triangles BCD, ACD, and ABD, respectively, and let h_A, h_B, and h_C be the lengths of the corresponding altitudes in the tetrahedron $ABCD$. Then $xS_A + yS_B + zS_C = 3V$, and the arithmetic mean–geometric mean inequality gives

$$xyz = \frac{1}{S_A S_B S_C}(xS_A)(yS_B)(zS_C)$$

$$\leq \frac{1}{S_A S_B S_C}\left(\frac{xS_A + yS_B + zS_C}{3}\right)^3 = \frac{V^3}{S_A S_B S_C}.$$

Equality holds when $xS_A = yS_B = zS_C = V$. Since $h_A S_A = h_B S_B = h_C S_C = 3V$, the latter is equivalent to

(1)
$$\frac{x}{h_A} = \frac{y}{h_B} = \frac{z}{h_C} = \frac{1}{3}.$$

It remains to describe the points M in $\triangle ABC$ for which (1) holds. Let A', B', and C' be the points where the lines AM, BM, and CM intersect the sides BC, AC, and AB, respectively. Then

$$\frac{x}{h_A} = \frac{MA'}{AA'}, \quad \frac{y}{h_B} = \frac{MB'}{BB'}, \quad \frac{z}{h_C} = \frac{MC'}{CC'},$$

and (1) is equivalent to $\frac{MA'}{AA'} = \frac{MB'}{BB'} = \frac{MC'}{CC'} = \frac{1}{3}$. The latter holds only when M is the centroid of $\triangle ABC$.

·**1.2.32** Draw a plane through M parallel to the plane OAB and let C_1 be the intersection point of this plane and OC. Denote by z the ratio of the distance from M to the plane OAB and the distance from C to OAB. Then $\frac{OC_1}{OC} = z$, i.e., $OC = \frac{OC_1}{z}$. Using similar notation, one gets $OA = \frac{OA_1}{x}$ and $OB = \frac{OB_1}{y}$. Therefore

$$OA^p \cdot OB^q \cdot OC^r = \frac{OA_1^p \cdot OB_1^q \cdot OC_1^r}{x^p y^q z^r}.$$

Since the segments OA_1, OB_1, and OC_1 do not depend on the plane through M, the right-hand side of the above equality is a minimum when the product $x^p y^q z^r$ is a maximum. Notice that $x + y + z = 1$, so the arithmetic mean–geometric mean inequality gives

$$\left(\frac{x}{p}\right)^p \left(\frac{y}{q}\right)^q \left(\frac{z}{r}\right)^r \le \left(\frac{p\left(\frac{x}{p}\right) + q\left(\frac{y}{q}\right) + r\left(\frac{z}{r}\right)}{p+q+r}\right)^{p+q+r} = \frac{1}{(p+q+r)^{p+q+r}}.$$

Equality holds when $x = \frac{p}{p+q+r}$, $y = \frac{q}{p+q+r}$, and $z = \frac{r}{p+q+r}$. This means that the plane α should be drawn in such a way that the barycentric coordinates of M in $\triangle ABC$ are $\left(\frac{p}{p+q+r}, \frac{q}{p+q+r}, \frac{r}{p+q+r}\right)$. The latter means that M is the intersection point of the lines AA_2, BB_2, and CC_2, where A_2, B_2, and C_2 divide the sides BC, CA, and AB into ratios $r : q$, $p : r$, and $q : p$, respectively.

1.2.33 Let x be the altitude of the part of the parallelepiped that is in the water. Then the volume of the water expelled is $V = abx$. The plane of the base of the parallelepiped cuts the container along a disk with radius $r = \sqrt{R^2 - x^2}$, circumscribed about a rectangle with sides a and b. Thus, $a^2 + b^2 = 4r^2 = 4(R^2 - x^2)$, so $x = \frac{1}{2}\sqrt{4R^2 - a^2 - b^2}$. Hence $V = \frac{ab}{2}\sqrt{4R^2 - a^2 - b^2}$. It then follows by the arithmetic mean–geometric mean inequality that

$$4V^2 = a^2 b^2 (4R^2 - a^2 - b^2) \le \left(\frac{a^2 + b^2 + 4R^2 - a^2 - b^2}{3}\right)^3 = \left(\frac{4}{3}R^2\right)^3,$$

where equality holds when $a = b = \frac{2R\sqrt{3}}{3}$. This is the case when a maximum amount of water will be expelled from the container.

4.3 Employing Calculus

1.3.7 Let $ABCD$ be the given quadrilateral, whose diagonals meet at O. For an arbitrary parallelogram $EFGH$ satisfying the conditions of the problem statement, set $AE = xAB$, where $0 < x < 1$. Then $EH = xBD$ and $EF = (1 - x)AC$. It is also clear that $\sin \angle FEH = \sin \alpha$, where α is the angle between the diagonals AC and BD. Hence

$$[EFGH] = EH \cdot EF \sin \alpha = x(1 - x)AC.BD \sin \alpha,$$

and because $[ABCD] = \frac{1}{2}AC \cdot BD \sin \alpha$, we obtain $[EFGH] = 2x(1 - x)S$.

The maximum value of the quadratic function $x(1 - x)$ in the interval $(0, 1)$ is $1/4$, and it is attained at $x = 1/2$. Therefore the maximum value of the area of the parallelogram $EFGH$ is $S/2$, and it is attained when its vertices are the midpoints of the sides of the given quadrilateral.

1.3.8 Let g and h be the given lines, and let ℓ be the line through A perpendicular to g (Fig. 113).

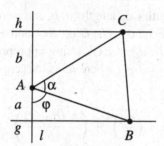

Figure 113.

Let $\angle BAC = \alpha$, where B is a point on g and C a point on h. Then B and C lie on the same side of ℓ. Denote by φ the angle between BA and ℓ. Then the angle between CA and ℓ is $180° - \alpha - \varphi$, which implies $AB = \frac{a}{\cos \varphi}$, $CA = -\frac{b}{\cos(\alpha+\varphi)}$, and therefore

$$[ABC] = -\frac{ab \sin \alpha}{2 \cos \varphi \cdot \cos(\alpha + \varphi)} = -\frac{ab \sin \alpha}{\cos \alpha + \cos(\alpha + 2\varphi)}.$$

It is now clear that $[ABC]$ is a maximum when $\alpha + 2\varphi = 180°$, i.e., when $\varphi = 90° - \frac{\alpha}{2}$. In this case $[ABC] = ab \cdot \cot \frac{\alpha}{2}$.

1.3.9 One observes immediately that the vertices of the required triangle must lie on the sides of the hexagon. Let AB be parallel to a side PQ of the hexagon. We may assume that PQ is the side of the hexagon closest to AB with this property. Then clearly C must lie on the opposite side MN of the hexagon (Fig. 114).

Set $a = PQ$, and let $2h$ be the distance from PQ to MN. Then $h = \frac{a\sqrt{3}}{2}$. Denote by x the distance between AB and PQ. Then $0 \le x \le h$ and the distance y from C to AB is $y = 2h - x$. On the other hand, using similar triangles, it

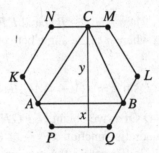

Figure 114.

follows that $AB = \frac{a(x+h)}{h}$. Hence $[ABC] = \frac{a(x+h)(2h-x)}{2h}$. The quadratic function $f(x) = (x+h)(2h-x)$ has a maximum when $x = \frac{h}{2}$. Thus $[ABC] \leq \frac{9ah}{8}$, where equality holds when $x = \frac{h}{2}$. The position of C on MN can be arbitrary.

1.3.10 Let T be a triangle ABC with side lengths a, b, and c and $EFGH$ a rectangle inscribed in T, where E and F lie on AB, G on BC, and H on AC (Fig. 115). Set $x = CH$, $u = HG$, $v = HE$, $d_c = EG$. Using appropriate pairs of similar triangles, one gets $\frac{u}{c} = \frac{x}{b}$ and $\frac{v}{h_c} = \frac{b-x}{b}$, where h_c is the length of the altitude of $\triangle ABC$ through C. Then

$$d_c^2 = u^2 + v^2 = \left(\frac{cx}{b}\right)^2 + \left(\frac{h_c(b-x)}{b}\right)^2.$$

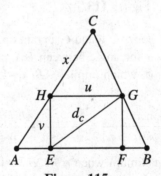

Figure 115.

The right-hand side of the above identity is a quadratic function of x that has a minimum value $d_c^2 = \frac{c^2 h_c^2}{c^2 + h_c^2} = \frac{4S^2(T)}{c^2 + h_c^2}$. Similarly, if two vertices of the rectangle lie on BC or CA, we get that the respective minimum value for d_a^2 equals $\frac{4S^2(T)}{a^2 + h_a^2}$ and for d_b^2 equals $\frac{4S^2(T)}{b^2 + h_b^2}$. If $a \leq b$, it follows from $a^2 + h_a^2 = a^2 + b^2 \sin^2 \gamma$ and $b^2 + h_b^2 = b^2 + a^2 \sin^2 \gamma$ that $a^2 + h_a^2 \geq b^2 + h_b^2$.

Suppose now that $a \leq b \leq c$. Then $d^2(T) = d_c^2 = \frac{4S^2(T)}{c^2 + h_c^2}$ and we get that

$$\frac{d^2(T)}{S(T)} = \frac{2ch_c}{c^2 + h_c^2} = \frac{2}{x + \frac{1}{x}},$$

where $x = \frac{h_c}{c}$. Since

(1)
$$\frac{h_c}{c} = \frac{b \sin A}{c} \leq \frac{c \sin 60°}{c} = \frac{\sqrt{3}}{2} < 1$$

and the function $f(x) = x + \frac{1}{x}$ is decreasing for $x \in (0, 1)$, we conclude that

$$\frac{d^2(T)}{S(T)} \leq \frac{2}{f(\sqrt{3}/2)} = \frac{4\sqrt{3}}{7}.$$

Equality holds precisely when $\frac{h_c}{c} = \frac{\sqrt{3}}{2}$. Now (1) implies that $c = b$ and $A = 60°$, i.e., ABC is an equilateral triangle.

1.3.11 Let $a = AD$, $\alpha = \angle DFE$, and $S = [EFD]$ (Fig. 23). Then $S = \frac{1}{2}[DED'F] = \frac{1}{4}EF \cdot DD'$. Since $\angle AED' = 2\alpha$, setting $x = DE$, we have $ED' = x$, and so $EA = x \cos 2\alpha$. This implies $a = x + x \cos 2\alpha$ and $x = \frac{a}{1 + \cos 2\alpha}$. Then $EF = \frac{x}{\sin \alpha} = \frac{a}{\sin \alpha (1 + \cos 2\alpha)}$, $DD' = \frac{a}{\cos \alpha}$, and therefore

$$S = \frac{a^2}{2} \cdot \frac{1}{\sin 2\alpha \, (1 + \cos 2\alpha)} = \frac{a^2}{8 \sin \alpha \cos^3 \alpha}.$$

Hence we have to find the maximum of the function $f(\alpha) = \sin \alpha \cos^3 \alpha$ for $\alpha \in (0°, 90°)$. Since

$$f'(\alpha) = \cos^4 \alpha - 3 \sin^2 \alpha \cos^2 \alpha = \cos^2 \alpha (1 - 4 \sin^2 \alpha),$$

it is easy to see that the maximum of $f(\alpha)$ is attained for $\alpha = 30°$ and is equal to $f(30°) = \frac{3\sqrt{3}}{16}$. Thus the minimum of S is equal to $\frac{2\sqrt{3}a^2}{9}$.

1.3.12 Let MN be the diameter of the half-disk and let $ABCD$ be an arbitrary quadrilateral inscribed in the half-disk. We leave it to the reader to observe that it is enough to consider the case $A = M$ and $B = N$ (Fig. 116).

For a fixed point C it is clear that $[ACD]$ is a maximum when D is the midpoint of the arc $\overset{\frown}{AC}$. So we may assume that $\angle AOD = \angle COD = \alpha$, $0° < \alpha < 90°$. Then $[ABCD] = \frac{R^2}{2}(2 \sin \alpha + \sin 2\alpha)$, and it is easy to see that the maximum of $[ABCD]$ is $\frac{3R^2\sqrt{3}}{4}$. It is attained only when $\alpha = 60°$, i.e., when C and D divide the semicircle into three equal parts.

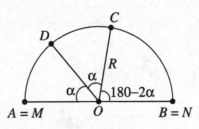

Figure 116.

1.3.13 Suppose that the center O of the disk lies outside the quadrilateral. Then there is a diameter of the disk such that the quadrilateral lies inside one of the half-disks determined by that diameter. Hence it follows from Problem 1.3.12 that the area of the quadrilateral is less than or equal to $\frac{3\sqrt{3}}{4}$, a contradiction.

1.3.14 Let M be a point on the circumcircle k of $\triangle ABC$ and set $t(M) = AM + BM + CM$. If M lies on one of the arcs $\overset{\frown}{AC}$ and $\overset{\frown}{BC}$, then for the reflection M' of M in the line AB we get $t(M') > t(M)$ since $AM = AM'$, $BM = BM'$, and $CM \leq CM'$. That is why it is enough to consider only points M on k such that MC intersects AB (Fig. 117).

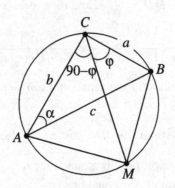

Figure 117.

Set $\varphi = \angle MCB$. It follows from the law of sines that $BM = c \sin \varphi$, $AM = c \sin(90° - \varphi) = c \cos \varphi$, and $CM = c \sin(\alpha + \varphi) = c \sin \alpha \cos \varphi + c \cos \alpha \sin \varphi = a \cos \varphi + b \sin \varphi$. Hence $t(M) = (a + c) \cos \varphi + (b + c) \sin \varphi$. We leave it to the reader to check that the function of $\varphi \in [0, 90°]$ obtained in this way achieves a maximum when $\tan \varphi = \frac{b+c}{a+c}$ and in that case $t(M) = \sqrt{(a + c)^2 + (b + c)^2}$.

Remark. The problem can also be solved using the Cauchy–Schwarz inequality.

1.3.15

(a) Let AB be an arbitrary chord in k. We are going to find all points C on k such that $AC^2 + BC^2$ is a maximum. We may assume that C lies on the larger of the arcs $\overset{\frown}{AB}$; then $\angle ACB = \alpha$ is constant and $0 \leq \alpha \leq 90°$. Setting $\varphi = \angle BAC$, we have $\angle ABC = 180° - \alpha - \varphi$ and by the law of sines we get

$$AC^2 + BC^2 = 4[\sin^2 \varphi + \sin^2(\alpha + \varphi)] = 2[2 - \cos 2\varphi - \cos 2(\alpha + \varphi)]$$
$$= 2[2 - 2\cos(\alpha + 2\varphi) \cdot \cos \alpha] \leq 4(1 + \cos \alpha),$$

where equality holds if and only if $\alpha + 2\varphi = 180°$, i.e., when C is the midpoint of the arc $\overset{\frown}{AB}$.

It remains to find the maximum of $s(M)$ when M is an isosceles triangle with an acute angle α at its top vertex. In this case

$$s(M) = 4\left(2\cos^2 \frac{\alpha}{2} + \sin^2 \alpha\right) = 4(1 + \cos \alpha + \sin^2 \alpha) = 4(2 + \cos \alpha - \cos^2 \alpha).$$

The quadratic function $2 + t - t^2$ achieves its maximum $\frac{9}{4}$ when $t = \frac{1}{2}$. Hence $s(M) \leq 9$, where equality holds when $\cos \alpha = \frac{1}{2}$, i.e., when $\alpha = 60°$ and M is an equilateral triangle.

(b) Let $n > 3$ and let M be an n-gon $A_1 A_2 \ldots A_n$ inscribed in k. There is an angle of M that is at least $90°$, e.g., assume that $\angle A_{n-1} A_n A_1 \geq 90°$. Then $A_1 A_n^2 + A_{n-1} A_n^2 \leq A_1 A_{n-1}^2$, so for the $(n-1)$-gon $M' = A_1 \ldots A_{n-1}$ we get $s(M) \leq s(M')$. Similarly (if $n - 1 > 3$), one constructs an $(n-2)$-gon M'' inscribed in k with $s(M') \leq s(M'')$, etc. One ends up with a triangle N inscribed in k such that $s(M) \leq s(N)$. From part (a), $s(N) \leq 9$ with equality only when N is equilateral. In the latter case we have $s(M) < s(N)$, so $s(M) < 9$.

Next, assume that $n \geq 4$. We will show that for any $\epsilon > 0$ there exists an n-gon M inscribed in k such that $s(M) > 9 - \epsilon$. Let $A_1 A_2 A_3$ be an equilateral triangle inscribed in k. Choose arbitrary points $A_4, A_5, \ldots, A_n$ on the arc $\overset{\frown}{A_3 A_1}$ such that $A_1 A_2 \ldots A_n$ is a convex n-gon (inscribed in k) and $A_1 A_n^2 > A_1 A_3^2 - \epsilon$. Then

$$s(M) = A_1 A_2^2 + A_2 A_3^2 + \sum_{i=3}^{n-1} A_i A_{i+1}^2 + A_n A_1^2$$
$$> 9 - (A_1 A_3^2 - A_1 A_n^2) > 9 - \epsilon,$$

and statement (b) is proved.

1.3.16 One has to consider two cases.

Case 1. Let $n = 2m$. Then the last circle is tangent to the first. Moreover, if x is the radius of the first circle, then we will have m circles of radius x and m of radius $a - x$. It is now easy to see that if S is the area of the n-gon and S_1 the area of the part of the n-gon outside the circles, then

$$S_1 = S - 2(m - 1)\pi [x^2 + (a - x)^2].$$

The maximum of this function when $x \in (0, a)$ is attained at $x = \frac{a}{2}$.

Case 2. Let $n = 2m+1$. If $x > \frac{a}{2}$, then the first and the last circles intersect. Thus, if we replace every circle with radius x (resp. $a-x$) by a concentric circle with radius $a - x$ (resp. x), then for the new set of circles the area S_1 will be larger. Hence we may assume that $0 \le x \le \frac{a}{2}$. Then

$$S_1 = S - \frac{(2m - 1)\pi}{2(2m + 1)} [(m + 1)x^2 + m(a - x)^2],$$

and the maximum of this function is attained at $x = \frac{ma}{2m+1}$.

1.3.17 *Hint.* There exists a side $B_1 B_2$ of the $(n + 1)$-gon that lies entirely on a side $A_1 A_2$ of the n-gon. Let $b = B_1 B_2$ and $a = A_1 A_2$. Show that $b = \frac{n}{n+1}a$. Then for $x = A_1 B_1$ we have $0 \le x \le \frac{a}{n+1}$ and the area S of the $(n + 1)$-gon is given by

$$S(x) = \frac{\sin \varphi}{2} \sum_{i=1}^{n} \left(\frac{i - 1}{n + 1} a + x \right) \left(\frac{n - i + 1}{n + 1} a - x \right),$$

where $\varphi = \angle A_1 A_2 A_3$. Thus $S(x)$ is a quadratic function of x. Show that $S(x)$ is minimal when $x = 0$ or $x = \frac{a}{n+1}$, and $S(x)$ is maximal when $x = \frac{a}{2(n+1)}$.

1.3.18 We may assume that the given circle k has radius 1 and C belongs to the larger arc $\overset{\frown}{AB}$ of k. Then $\angle ACB = \alpha$ is a constant and $0 \le \alpha \le 90°$. Set $\varphi = \angle BAC$. Then the law of sines gives $AC = 2 \sin \varphi$, $BC = 2 \sin(\alpha + \varphi)$.

(a) We have

$$AC + BC = 2(\sin \varphi + \sin(\alpha + \varphi)) = 4 \sin \frac{\alpha + 2\varphi}{2} \cos \frac{\alpha}{2} \le 4 \cos \frac{\alpha}{2}.$$

Hence the maximum of $AC + BC$ is attained when $\alpha + 2\varphi = 180°$, i.e., when C is the midpoint of the arc $\overset{\frown}{AB}$.

(b) It follows from the solution of Problem 1.3.15 (a) that the maximum of $AC^2 + BC^2$ is attained when C is the midpoint of the arc $\overset{\frown}{AB}$.

(c) We have

$$AC^3 + BC^3 = 8[\sin^3 \varphi + \sin^3(\alpha + \varphi)]$$

$$= 8[\sin \varphi + \sin(\alpha + \varphi)]$$

$$\times [\sin^2 \varphi - \sin \varphi \cdot \sin(\alpha + \varphi) + \sin^2(\alpha + \varphi)]$$

$$= 8 \cos \frac{\alpha}{2} \sin\left(\varphi + \frac{\alpha}{2}\right)$$

$$\times [2 - \cos \alpha + \cos(2\varphi + \alpha) - 2\cos(2\varphi + \alpha)\cos \alpha].$$

Set $t = \sin\left(\frac{\alpha}{2} + \varphi\right)$. Then $0 \leq t \leq 1$ and

$$\cos(\alpha + 2\varphi) = 1 - 2\sin^2\left(\frac{\alpha}{2} + \varphi\right) = 1 - 2t^2.$$

Therefore

$$AC^3 + BC^3 = 8 \cos \frac{\alpha}{2} t[2 - \cos \alpha + (1 - 2\cos \alpha)(1 - 2t^2)]$$

$$= 8 \cos \frac{\alpha}{2} [3(1 - \cos \alpha)t - 2(1 - 2\cos \alpha)t^3]$$

$$= 8 \cos \frac{\alpha}{2} \cdot g(t).$$

For the function $g(t)$ we have $g'(t) = 3(1 - \cos \alpha) - 6(1 - 2\cos \alpha)t^2$.

Case 1. $0 \leq \alpha \leq 60°$. Then $1 \geq \cos \alpha \geq \frac{1}{2}$ and $g'(t) > 0$ for all t, which means that $g(t)$ is an increasing function of t. Since $t \leq 1$, it follows that $AC^3 + BC^3$ is a maximum when $t = 1$, i.e., when $\frac{\alpha}{2} + \varphi = 90°$.

In this case C is the midpoint of the arc $\overset{\frown}{AB}$ and $AC^3 + BC^3 = 8 \cos \frac{\alpha}{2} (1 + \cos \alpha)$.

Case 2. $60° < \alpha \leq 90°$. Then $0 \leq \cos \alpha < \frac{1}{2}$ and $1 - 2\cos \alpha > 0$. In this case $g'(t) = 0$ when $t^2 = \frac{1 - \cos \alpha}{2(1 - 2\cos \alpha)}$.

(a) $\frac{1}{3} \leq \cos \alpha < \frac{1}{2}$. Then $\frac{1 - \cos \alpha}{2(1 - 2\cos \alpha)} \geq 1$, which means that $g'(t) > 0$ for $t \in [0, 1)$. Thus $g(t)$ is again a strictly increasing function in $[0, 1]$ and achieves its maximum at $t = 1$, i.e., when C is the midpoint of the arc $\overset{\frown}{AB}$.

(b) $0 \le \cos \alpha < \frac{1}{3}$. Now we have $0 < \frac{1-\cos\alpha}{2(1-2\cos\alpha)} < 1$, so

$$t_0 = \sqrt{\frac{1 - \cos \alpha}{2(1 - 2 \cos \alpha)}} \in (0, 1).$$

Clearly $g'(t_0) = 0$ and $g(t_0)$ is the maximal value of $g(t)$ for $t \in [0, 1]$. We have

$$g(t_0) = \sqrt{\frac{1 - \cos \alpha}{2(1 - 2 \cos \alpha)}}[3(1 - \cos \alpha) - (1 - \cos \alpha)]$$

$$= \frac{\sqrt{2}(1 - \cos \alpha)^{3/2}}{\sqrt{1 - 2 \cos \alpha}} = \frac{4 \sin^3 \frac{\alpha}{2}}{\sqrt{1 - 2 \cos \alpha}},$$

so in this case the maximum of $AC^3 + BC^3$ is equal to

$$\frac{32 \sin^3 \frac{\alpha}{2} \cos \frac{\alpha}{2}}{\sqrt{1 - 2 \cos \alpha}} = \frac{8 \sin \alpha (1 - \cos \alpha)}{\sqrt{1 - 2 \cos \alpha}},$$

and it is attained when

$$\sin \left(\frac{\alpha}{2} + \varphi\right) = t_0 = \sqrt{\frac{1 - \cos \alpha}{2(1 - 2 \cos \alpha)}} \in (0, 1).$$

Notice that $t_0 = \frac{\sin \frac{\alpha}{2}}{\sqrt{1-2\cos\alpha}} > \sin \frac{\alpha}{2}$, so $t_0 = \sin \beta$ for some β with $\frac{\alpha}{2} < \beta < 90°$. The value of φ for which $AC^3 + BC^3$ achieves a maximum is now given by $\frac{\alpha}{2} + \varphi = \beta$ or $\frac{\alpha}{2} + \varphi = 180° - \beta$, i.e., when $\varphi = \beta - \frac{\alpha}{2}$ or $\varphi = 180° - \beta - \frac{\alpha}{2}$.

1.3.19 It is easy to see that it is enough to consider only points X lying in the half-plane determined by ℓ that contains A and B. Let the distance from A to ℓ be a and let that from B to ℓ be b. We may assume that $a \le b$. Consider the coordinate system Oxy in the plane such that the x-axis coincides with ℓ and the positive y-axis contains A. Then A has coordinates $(0, a)$ and B has coordinates (d, b). We may assume that $d \ge 0$. Notice that if X is a point in the upper half-plane such that the line ℓ' passing through X and parallel to ℓ intersects the ray issuing from A and passing through B, then $t(A) < t(X)$. That is why it is enough to consider the case that the distance from ℓ' to ℓ does not exceed a, i.e., the case that X has coordinates (x, y) with $0 \le y \le a$.

Fix $y \in [0, a]$ and denote by ℓ' the horizontal line in the upper half-plane whose distance to ℓ is y. If B'_y is the reflection of B in ℓ', it follows from Problem 1.1.1 that

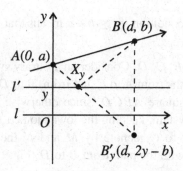

Figure 118.

for $X \in \ell'$ the sum $AX + XB$ is minimal when X coincides with the intersection point X_y of ℓ' and AB'_y (Fig. 118).

Thus, for $X \in \ell'$ the sum $t(X)$ is minimal when $X = X_y$ and $t(X_y) = y + AB'_y$. Since the coordinates of B'_y are $(d, 2y - b)$, it follows that $t(X_y) = y + \sqrt{d^2 + (a + b - 2y)^2}$. It remains to find the minimum of the function $f(y) = y + \sqrt{d^2 + (a + b - 2y)^2}$ on $[0, a]$. We have

$$f'(y) = 1 - \frac{2(a + b - 2y)}{\sqrt{d^2 + (a + b - 2y)^2}}$$

$$= \frac{d^2 - 3(a + b - 2y)^2}{\sqrt{d^2 + (a + b - 2y)^2} \cdot [\sqrt{d^2 + (a + b - 2y)^2} + 2(a + b - 2y)]},$$

and $f'(y) = 0$ only if $y = y_0 = \frac{1}{2}\left(a + b - \frac{d}{\sqrt{3}}\right)$. Depending on the position of y_0, there are three possible cases.

Case 1. $a + b \le \frac{d}{\sqrt{3}}$. Then $y_0 \le 0$, so $f'(y) > 0$ for all $y \in (0, a]$ and $f(y)$ is strictly increasing on this interval. Thus $f(y)$ is minimal for $y = 0$. In this case $t(X)$ is minimal when X coincides with the point $X_0 \in \ell$ for which the segments AX_0 and BX_0 make equal angles with ℓ.

Case 2. $\frac{d}{\sqrt{3}} \le b - a$. Then $y_0 \ge a$, so $f'(y) < 0$ for $y \in [0, a)$ and $f(y)$ is strictly decreasing on this interval. Thus its minimal value is $f(a)$. In other words, $t(X)$ is minimal when $X = A$.

Case 3. $b - a < \frac{d}{\sqrt{3}} < a + b$. Then $y_0 \in (0, a)$, so $f(y)$ has a minimum at $y = y_0$. Thus, $t(X)$ is minimal when $X = \left(\frac{d}{2} - \frac{\sqrt{3}}{2}(b - a), \frac{a+b}{2} - \frac{d}{2\sqrt{3}}\right)$. It is not difficult to check that in this case $\angle AXB = 120°$.

It should be mentioned that the condition $\frac{d}{\sqrt{3}} \leq b - a$ means that the angle between AB and ℓ is not less than $30°$.

1.3.20 Let the four towns be A, B, C, D. Consider an arbitrary system of highways joining them. Then there are paths from A to C and from B to D. We may assume that these paths lie inside the square $ABCD$, since otherwise one could clearly shorten the total length of the system, keeping the towns joined.

Following the path from A to C, denote by M and N the first and the last intersection points of this path with the path from B to D (Fig. 119).

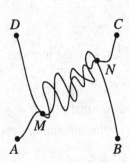
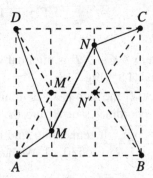

Figure 119.   **Figure 120.**

We can shorten the total length of the given system of highways by replacing it with the system consisting of the five line segments

$$AM, \quad DM, \quad MN, \quad BN, \quad CN.$$

Draw the parallels through M and N to AD and BC. Then choose on these parallels points M' and N', respectively, that are equidistant from the sides AB and CD (Fig. 120). It follows from Heron's problem (Problem 1.1.1) that

$$AM + DM \geq AM' + DM' \quad \text{and} \quad BN + CN \geq BN' + CN'.$$

It is also clear that $MN \geq M'N'$, because $M'N'$ is the distance between the parallels considered above. Adding these inequalities gives

$$AM + DM + MN + BN + CN \geq AM' + DM' + M'N' + BN' + CN'.$$

Thus we have reduced our problem to the following:

Let E and F be the midpoints of the sides AD and BC of the square $ABCD$. Find points M and N on the line segment EF such that $AM + DM + MN + BN + CN$ is a minimum.

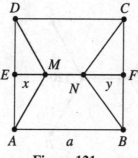

Figure 121.

Denote the side length of $ABCD$ by a, and let $EM = x$, $FN = y$, where $0 \le x \le a$, $0 \le y \le a - x$ (Fig. 121).

Then

$$AM = MD = \sqrt{x^2 + \frac{a^2}{4}}, \quad MN = a - x - y, \quad BN = CN = \sqrt{y^2 + \frac{a^2}{4}}.$$

Hence we have to determine the minimum of the function

$$F(x, y) = 2\sqrt{x^2 + \frac{a^2}{4}} + a - x - y + 2\sqrt{y^2 + \frac{a^2}{4}}$$

for $0 \le x \le a$, $0 \le y \le a - x$. Consider the function $f(x) = 2\sqrt{x^2 + \frac{a^2}{4}} - x$. Its derivative

$$f'(x) = \frac{2x}{\sqrt{x^2 + \frac{a^2}{4}}} - 1 = \frac{2}{\sqrt{1 + \frac{a^2}{4x^2}}} - 1$$

is strictly increasing in $(0, +\infty)$, and $f'(x) = 0$ for $x = a/(2\sqrt{3})$. It follows that the minimum value of $f(x)$ in the interval $[0, a]$ is attained at $x = a/(2\sqrt{3})$ and is equal to $a\sqrt{3}/2$.

Since $F(x, y) = f(x) + f(y) + a$, one easily infers from here that the minimum of the function $F(x, y)$ is attained at $x = a/(2\sqrt{3})$, $y = a/(2\sqrt{3})$. This minimum is equal to $a(1 + \sqrt{3})$. Hence the solution of our problem is given (up to symmetry) by the system of highways shown in Figure 122.

Remark. After reducing the problem to finding the minimum of $AM + DM + MN + NB + NC$ (Fig. 120) we may proceed in a shorter way. Let P and Q be the points outside $ABCD$ such that ADP and BCQ are equilateral triangles. Then by Pompeiu's theorem (Problem 1.1.6) it follows that $AM + DM \ge PM$ and $BN + CN \ge QN$. Hence

$$AM + DM + MN + NB + NC \ge PM + MN + NQ \ge PQ.$$

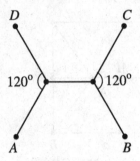

Figure 122.

Thus the desired minimum is equal to PQ and it is attained for the system of highways shown in Fig. 122.

1.3.21 Let a plane through the vertex C of the cone intersect the circle of its base at points A and B. Let R be the radius of the base, $AC = BC = \ell$, and set $AB = 2x$, $0 < x \le R$. Then $[ABC] = x\sqrt{\ell^2 - x^2} = \sqrt{x^2(\ell^2 - x^2)}$, and we have to find the maximum of the quadratic function $f(t) = t(\ell^2 - t)$ on the interval $(0, R^2]$. If $\ell^2 \le 2R^2$ then the maximum of $f(t)$ is attained at $t = \frac{\ell^2}{2}$ and is equal to $\frac{\ell^4}{4}$. In this case $AB = \ell\sqrt{2}$, $[ABC] = \frac{\ell^2}{2}$, and we note that $\angle ACB = 90°$. If $\ell^2 \ge 2R^2$ then the maximum of $f(t)$ is attained at $t = R^2$ and is equal to $R^2(\ell^2 - R^2)$. In this case $AB = 2R$ (i.e., AB is a diameter of the base) and $[ABC] = R\sqrt{\ell^2 - R^2}$.

1.3.22 For any point X in α set $x = PX$ and $\varphi = \angle XPQ$. Then

$$d(X) = \frac{x + PQ}{\sqrt{x^2 + PQ^2 - 2xPQ\cos\varphi}},$$

and for a fixed x this is a maximum when $\cos\varphi$ is a maximum. This happens when $PQ \perp \alpha$ (then $\varphi = 90°$ for any $X \in \alpha$), or when X lies on the orthogonal projection of the ray r issuing from P and passing through Q onto α (Fig. 123).

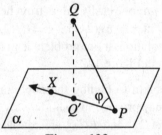

Figure 123.

In what follows we consider only such points X. Let φ_0 be the angle between the ray r and α and let $a = PQ$. It is not difficult to check that the function

$f(x) = \frac{(x+a)^2}{x^2+a^2-2ax\cos\varphi_0}$ achieves its maximum precisely when $x = a$. Thus, $d(X)$ is a maximum when $PX = PQ$.

1.3.23

(a) Let the edge of the cube be of length 1, $BM = x$, $0 \leq x \leq 1$, and $\angle B_1 M C_1 = \varphi$ (Fig. 124). Then

$$B_1 M = \sqrt{1+x^2}, \quad C_1 M = \sqrt{B_1 M^2 + B_1 C_1^2} = \sqrt{2+x^2}$$

(since $\angle C_1 B_1 M = 90°$) and we obtain

$$\cos\varphi = \frac{B_1 M}{C_1 M} = \sqrt{\frac{1+x^2}{2+x^2}}.$$

Hence $\cos\varphi \geq 1/\sqrt{2}$, because $(1+x^2)/(2+x^2) \geq 1/2$, with equality only if $x = 0$. Thus the angle $\angle B_1 M C_1$ is a maximum if M coincides with B, and in this case $\angle B_1 M C_1 = 45°$.

(b) Let $AM = x$, $0 \leq x \leq 1$, and $\angle A_1 M C_1 = \varphi$. Then $A_1 M = \sqrt{1+x^2}$, $C_1 M = \sqrt{2+(1-x)^2}$, $A_1 C_1 = \sqrt{2}$ (Fig. 124).

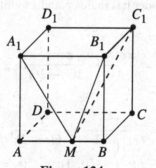

Figure 124.

The law of cosines for $\triangle A_1 M C_1$ gives

$$\cos\varphi = \frac{A_1 M^2 + C_1 M^2 - A_1 C_1^2}{2A_1 M \cdot C_1 M} = \frac{x^2 - x + 1}{\sqrt{x^2+1}\sqrt{x^2-2x+3}}.$$

Since $x^2 - x + 1 > 0$ for all x, it suffices to find the minimum of the function

$$f(x) = \frac{(x^2 - x + 1)^2}{(x^2 + 1)(x^2 - 2x + 3)}.$$

on the interval $[0, 1]$. We have

$$f'(x) = \frac{2(x^2 - x + 1)(x^3 + 3x - 2)}{(x^2 + 1)^2(x^2 - 2x + 3)^2},$$

and therefore the sign of $f'(x)$ is determined by the sign of the function $g(x) = x^3 + 3x - 2$ on $[0, 1]$. Since $g(x)$ is strictly increasing ($g'(x) = 3x^2 + 3 > 0$) and also $g(0) = -2$, $g(1) = 2$, it follows from the intermediate value theorem that the equation $x^3 + 3x - 2 = 0$ has a unique solution $x_0 \in (0, 1)$. Hence the function $f(x)$ is decreasing on the interval $(0, x_0)$ and increasing on the interval $(x_0, 1)$. On the other hand, $f(0) = 1/3 > 1/4 = f(1)$. So the maximum of $f(x)$ on $[0, 1]$ is attained at $x = 0$ and is equal to $1/3$. Thus $\angle A_1 M C_1$ is a minimum when M coincides with A and in this case $\cos \varphi = \frac{1}{\sqrt{3}}$.

Remark.   The arguments above show that the minimum of the function $f(x)$ on $[0, 1]$ is attained at x_0. Hence $\angle A_1 M C_1$ is a maximum for the point M_0 on AB such that $A M_0 = x_0$. Note that $x_0 = \sqrt[3]{\sqrt{2} + 1} - \sqrt[3]{\sqrt{2} - 1}$.

1.3.24 Denote by r, x, and a the radius of the sphere, the altitude of the cone, and the radius of its base, respectively. Then $xa = r(a + \sqrt{a^2 + x^2})$ and we get

(1) $$a^2 = \frac{r^2 x}{x - 2r}, \quad x > 2r.$$

It is clear that the base of the cylinder has radius r and its altitude is $2r$.

(a) It follows from (1) that

$$V_1 = \frac{\pi a^2 x}{3} = \frac{\pi r^2 x^2}{3(x - 2r)}.$$

Since $V_2 = 2\pi r^3$, we obtain

$$\frac{V_1}{V_2} = \frac{x^2}{6r(x - 2r)} = \frac{t^2}{6(t - 2)},$$

where $t = x/r > 2$. Set $f(t) = t^2/(t - 2)$. Then $f'(t) = \frac{t(t-4)}{(t-2)^2}$, which shows that the function $f(t)$ is decreasing on the interval $(2, 4)$ and increasing on the interval $(4, +\infty)$. Hence $f(t)$ has a minimum at $t = 4$ and this minimum is equal to 8. Thus $V_1/V_2 \geq 4/3$.

(b) We have

$$S_1 = \frac{\pi r x(x - r)}{x - 2r} \quad \text{and} \quad S_2 = 4\pi r^2.$$

Hence

$$\frac{4S_1}{S_2} = \frac{t(t - 1)}{t - 2} = f(t),$$

where $t = x/r > 2$. Since

$$f'(t) = \frac{t^2 - 4t + 2}{(t - 2)^2},$$

the function $f(t)$ decreases on the interval $(2, 2 + \sqrt{2})$ and increases on the interval $(2 + \sqrt{2}, +\infty)$. Hence $f(t)$ has a minimum at $t = 2 + \sqrt{2}$, and this minimum equals $3 + 2\sqrt{2}$. Thus $4\frac{S_1}{S_2} \geq 3 + 2\sqrt{2}$.

1.3.25 *Answer.* If O_1 and O_2 are the centers of the spheres and R_1 and R_2 their radii, then the distance between the light source and O_1 must be $x = \frac{R_1\sqrt{R_1}}{R_1\sqrt{R_1} + R_2\sqrt{R_2}} O_1 O_2$.

4.4 The Method of Partial Variation

1.4.6 Fix an arbitrary point M on k. Let M' and M'' be the reflections of M in p and q, respectively (Fig. 125).

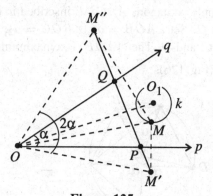

Figure 125.

We want to find points P on p and Q on q such that $\triangle MPQ$ has a minimal perimeter. It follows from Problem 1.1.16 that the solution is given by the intersection points P and Q of $M'M''$ with p and q, respectively. In this case the perimeter of $\triangle MPQ$ is $M'M''$, and moreover $M'M''$ is the base of an isosceles triangle $M'M''O$ with a constant angle at the vertex O. Thus, $M'M''$ is minimal when the side $OM' = OM'' = OM$ is minimal, i.e., when M is the intersection point of k and the segment OO_1, where O_1 is the center of k.

1.4.7

(a) Let ABC be a triangle inscribed in k, and C' the midpoint of the larger arc $\overset{\frown}{AB}$. Since the distance from C' to the line AB is not less than the distance from C to AB, it follows that $[ABC] \leq [ABC']$. Set $2\gamma = \angle AC'B$, $0 < \gamma \leq 45°$. Then the law of sines gives $AB = 2R \sin 2\gamma$, where R is the radius of k. The altitude through C' of $\triangle ABC'$ is equal to $\frac{AB \cot \gamma}{2}$ and therefore $[ABC'] = 4R^2 \sin \gamma \cos^3 \gamma$. Consider the function $f(\gamma) = \sin \gamma \cos^3 \gamma$. Then $f'(\gamma) = \cos^2 \gamma \, (1 - 4 \sin^2 \gamma)$, and it follows that the maximum of $f(\gamma)$ on the interval $(0, 45°]$ is attained at $\gamma = 30°$, i.e., when triangle ABC' is equilateral.

Thus, of all triangles inscribed in k the equilateral triangles have maximum area.

(b) Let $ABCD$ be a quadrilateral inscribed in k. Denote by α the angle between AC and BD. Then

$$[ABCD] = \frac{AC.BD.\sin \alpha}{2} \leq \frac{AC.BD}{2} \leq 2R^2.$$

Hence the maximum of $[ABCD]$ is $2R^2$ and it is attained if and only if $ABCD$ is a square.

(c) It is enough to consider only pentagons $ABCDE$ inscribed in the given circle and containing its center O. Set $\angle AOB = \alpha_1$, $\angle BOC = \alpha_2, \ldots, \angle EOA = \alpha_5$. Fix the points A, B, C, and D. Then $[ADE]$ is a maximum when E is the midpoint of the arc $\overarc{AD}$ (Fig. 126).

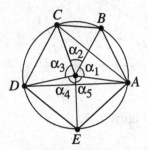

Figure 126.

Hence it is enough to consider only pentagons for which $\alpha_4 = \alpha_5 = \beta$. Similarly, we may assume that $\alpha_1 = \alpha_2 = \alpha$.

We then have $[ABCDE] = [A'B'C'D'E']$, where (Fig. 127) $\angle A'OB' = \angle A'OE' = \alpha$ and $\angle B'OC' = \angle D'OE' = \beta$, which implies that E' and D' are symmetric to the points B' and C', respectively, with respect to the line OA'. Fix for a moment A', C', and D'. The areas of triangles $A'B'C'$ and $A'D'E'$ are maximal when B' is the midpoint of $\overarc{A'C'}$ and E' is the midpoint of $\overarc{A'D'}$, i.e., when $\alpha = \beta$.

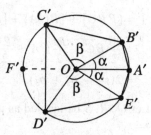

Figure 127.

Thus, it is enough to consider pentagons $ABCDE$ such that four of the central angles determined by their sides have the same measure α (then $180° \leq 4\alpha \leq 360°$, i.e., $45° \leq \alpha \leq 90°$). Then for the area $S(\alpha)$ of such a pentagon we have

$$S(\alpha) = \frac{R^2}{2}[4\sin\alpha + \sin(360° - 4\alpha)] = \frac{R^2}{2}[4\sin\alpha - \sin 4\alpha].$$

Hence

$$S'(\alpha) = 2R^2[\cos\alpha - \cos 4\alpha] = 4R^2 \sin\frac{5\alpha}{2}\sin\frac{3\alpha}{2}.$$

Since $\sin\frac{3\alpha}{2} > 0$ for all $\alpha \in [45°, 90°]$, it follows that $S'(\alpha) > 0$ for $45° \leq \alpha < 72°$ and $S'(\alpha) < 0$ for $72° < \alpha \leq 90°$. Thus, $S(\alpha)$ is a maximum when $\alpha = 72°$, in which case the pentagon is regular.

(d) *Hint.* As in (c), show that it is enough to consider only hexagons whose sides determine central angles $\alpha, \alpha, \alpha, \alpha, \beta, \beta$. The maximum area is achieved when the hexagon is regular.

1.4.8 It follows from Problem 1.4.2 that the area of $\triangle PQR$ is a maximum when P, Q, and R are vertices of the hexagon. Suppose that at least two of them are consecutive vertices. Then $\triangle PQR$ is contained in a quadrilateral formed by four consecutive vertices of the hexagon, and has area less than half the area of the hexagon. On the other hand, by symmetry, it is easy to see that

$$[ACE] = [BDF] = \frac{1}{2}[ABCDEF].$$

Hence the traingle PQR of maximum area is either ACE or BDF.

1.4.9

(a) For any point P inside $\triangle ABC$, denote its distances from BC, CA, AB by p_1, p_2, p_3, respectively. Note that

$$4\sqrt{3} = [ABC] = [PBC] + [PCA] + [PAB]$$

$$= \frac{p_1}{2}BC + \frac{p_2}{2}CA + \frac{p_3}{2}AB$$

$$= 2(p_1 + p_2 + p_3).$$

Hence $p_1 + p_2 + p_3 = 2\sqrt{3}$. Let Q, R, S be labeled as in Fig. 128.

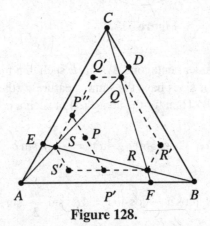

Figure 128.

Construct the hexagon $RR'QQ'SS'$ with QQ' and RS' parallel to BA, RR' and SQ' parallel to AC, and SS' and QR' parallel to BC. It is easy to see that this hexagon is situated symmetrically within $\triangle ABC$. It follows that

$$q_1 q_2 q_3 = q_1' q_2' q_3' = r_1 r_2 r_3 = r_1' r_2' r_3' = s_1 s_2 s_3 = s_1' s_2' s_3'.$$

For any point P inside $QQ'RR'SS'$, draw the line through it parallel to BC, cutting the perimeter of the hexagon at P' and P'', one of which may be P itself.

Then $p_1' = p_1$ and $p_2' + p_3' = p_2 + p_3$. Moreover,

$$|p_2' - p_3'| \geq |p_2 - p_3|.$$

Hence

$$0 \leq (p_2' - p_3')^2 - (p_2 - p_3)^2$$

$$= (p_2' + p_3')^2 - 4p_2' p_3' - (p_2 + p_3)^2 + 4p_2 p_3$$

$$= 4(p_2 p_3 - p_2' p_3').$$

Thus the product of the three distances does not increase if we replace P by P'. Now P' may already be a vertex of the hexagon. If not, it lies between two vertices, and the same argument shows that the product decreases if we replace P'

by a vertex. Restricting ourselves now to $\triangle QRS$, we see that the product is a minimum if P coincides with one of Q, R, and S.

(b) Note that triangles ASE, ACD, and CQD are similar. Hence $AS \cdot AD = AE \cdot AC = 4$ and $DQ \cdot AD = CD \cdot AE = 1$. By the law of cosines for $\triangle ACD$ we get that $AD = \sqrt{13}$ and therefore $AS : SQ : QD = 4 : 8 : 1$. Now the altitude of $\triangle ABC$ is $2\sqrt{3}$. Hence

$$s_1 = \frac{18\sqrt{3}}{13}, \quad s_2 = r_3 = \frac{2\sqrt{3}}{13}, \quad s_3 = 2\sqrt{3} - s_1 - s_2 = \frac{6\sqrt{3}}{13}.$$

Thus

$$s_1 s_2 s_3 = \frac{648\sqrt{3}}{2197}.$$

By (a), this is the minimum value of the desired product.

1.4.10 *Answer.* If A, B, and C are the given points, then the required line is the one passing along the largest side of $\triangle ABC$.

1.4.11 Fix the points A and D and let $\alpha = \overset{\frown}{AD} < 180°$. Since $AC \perp BD$, the center O of the circle lies in the quadrilateral $ABCD$ (Fig. 129).

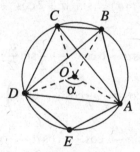

Figure 129.

Moreover, $\overset{\frown}{AD} + \overset{\frown}{BC} = 180°$, so $[AOD] = [BOC] = \frac{\sin \alpha}{2}$, which is constant (when α is fixed). Clearly $[AOB] \le \frac{1}{2}$, with equality when $\angle AOB = 90°$. The same applies to $[COD]$. Hence we may assume that $\angle AOB = \angle COD = 90°$. Moreover, $[ADE]$ is a maximum when E is the midpoint of $\overset{\frown}{AD}$.

Thus, it is enough to consider only pentagons $ABCDE$ for which $\angle AOB = \angle COD = 90°$ and E is the midpoint of $\overset{\frown}{AD}$. Then

$$[ABCDE] = 1 + \frac{\sin \alpha}{2} + \sin \frac{\alpha}{2}.$$

It is easy to see that the maximum of this function of α is attained when $\alpha = 120°$. Hence $[ABCDE]$ is a maximum when $\angle AOB = \angle COD = 90°$ and $\angle BOC = \angle DOE = \angle EOA = 60°$.

1.4.12 *Hint.* Follow the solutions of Problems 1.4.2 and 1.4.3.

Answer.

$$\frac{S}{n \sin \frac{360°}{n}} \left[(p - r) \sin \frac{k360°}{n} + r \sin \frac{(k+1)360°}{n} \right],$$

where $n = pk + r$, $0 \le r < p$.

1.4.13 Let ABC be an arbitrary triangle inscribed in the given circle k with center O. We may assume that BC is the smallest side of the triangle. Then $2\alpha = \angle BOC \le 120°$, i.e., $0 \le \alpha \le 60°$. It now follows from Case 1 in the solution of Problem 1.3.18 that when B and C are fixed, the sum $AB^3 + AC^3$ is a maximum when A is the midpoint of the larger arc $\overset{\frown}{BC}$.

In what follows we consider only isosceles triangles ABC ($AB = AC$) for which $\alpha = \angle BAC \le 60°$. Then

$$AB^3 + BC^3 + AC^3 = 8R^3 \left(\sin^3 \alpha + 2\cos^3 \frac{\alpha}{2} \right).$$

We need to investigate the function $f(\alpha) = \sin^3 \alpha + 2\cos^3 \frac{\alpha}{2}$ for $0 \le \alpha \le 60°$. We have

$$f'(\alpha) = 3 \sin^2 \alpha \cdot \cos \alpha - 6 \cos^2 \frac{\alpha}{2} \cdot \frac{1}{2} \sin \frac{\alpha}{2}$$

$$= 3 \left(\sin^2 \alpha \cdot \cos \alpha - \frac{1}{2} \cos \frac{\alpha}{2} \cdot \sin \alpha \right)$$

$$= 3 \sin \alpha \left(2 \sin \frac{\alpha}{2} \cdot \cos \frac{\alpha}{2} \cdot \cos \alpha - \frac{1}{2} \cos \frac{\alpha}{2} \right)$$

$$= \frac{3}{2} \sin \alpha \cdot \cos \frac{\alpha}{2} \left(4 \sin \frac{\alpha}{2} \cos \alpha - 1 \right)$$

$$= \frac{3}{2} \sin \alpha \cdot \cos \frac{\alpha}{2} \left[4 \sin \frac{\alpha}{2} - 8 \sin^3 \frac{\alpha}{2} - 1 \right].$$

When α runs over the interval $[0, 60°]$, $\sin \frac{\alpha}{2}$ runs over $[0, \frac{1}{2}]$. So, in order to investigate the sign of $f'(\alpha)$, it is enough to determine the sign of $g(t) = 4t - 8t^3 - 1$ for $t \in [0, \frac{1}{2}]$. One way to do this is to factorize $g(t)$ (this is not difficult since $g(1/2) = 0$). Here instead we deal with $g'(t)$. We have

$$g'(t) = 4 - 24t^2 = 4(1 - 6t^2),$$

so $g(t)$ is strcitly increasing on $[0, \frac{1}{\sqrt{6}}]$ and strictly decreasing in $[\frac{1}{\sqrt{6}}, \frac{1}{2}]$. Since $g(0) = -1 < 0$ and $g(1/2) = 0$ (Fig. 130), there exists a unique $t_0 \in (0, \frac{1}{\sqrt{6}})$ with $g(t_0) = 0$.

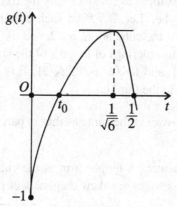

Figure 130.

Thus there exists a unique $\alpha_0 \in (0, 60°)$ such that $\sin \frac{\alpha_0}{2} = t_0$. Then $f'(\alpha) < 0$ for $\alpha \in [0, \alpha_0)$ and $f'(\alpha) > 0$ for $\alpha \in (\alpha_0, 60°]$ (Fig. 131).

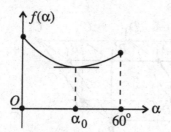

Figure 131.

It is now clear that $f(\alpha_0)$ is the minimum value of $f(\alpha)$, while its maximum is achieved either for $\alpha = 0$ or for $\alpha = 60°$. Since $f(0) = 2$ and $f(60°) = \frac{9\sqrt{3}}{8}$, we have $f(0) > f(60°)$. That is, the maximum of $f(\alpha)$ is achieved when $\alpha = 0$. This is equivalent to $B = C$. In this case (assuming A is diametrically opposite to $B = C$) we have $AB^3 + BC^3 + AC^3 = 16R^3$. For every nondegenerate triangle ABC this sum is strictly less than $16R^3$. However, the continuity of $f(\alpha)$ shows that it can be made arbitrarily close to $16R^3$.

1.4.14 *Hint.* Use the same argument as in the solution of Problem 1.4.2.

1.4.15 *Hint.* Use the argument from the solution of Problem 1.4.2.

1.4.16

(a) Let $ABCDA_1B_1C_1D_1$ be the given cube and let $a = AB$. We conclude from Problem 1.4.14 that it suffices to consider only the triangles with vertices among the vertices of the cube. Let MNP be such a triangle. The possible distances between vertices of the cube are a, $a\sqrt{2}$, and $a\sqrt{3}$. It is easy to see that the possible triples (up to ordering) of lengths of the sides of $\triangle MNP$ are $\{a, a, a\sqrt{2}\}$, $\{a, a\sqrt{2}, a\sqrt{3}\}$, and $\{a\sqrt{2}, a\sqrt{2}, a\sqrt{2}\}$. It is now easy to check that $[MNP]$ is a maximum in the third case.

(b) Use Problem 1.4.14. The answer is the same as that in part (a).

1.4.17 Use Problem 1.4.15. *Answer.* A tetrahedron in the cube has a maximum volume precisely when two of its edges are skew diagonals of parallel faces of the cube.

1.4.18 Let $ABCDA_1B_1C_1D_1$ be an arbitrary prism of volume V. Construct points A_1' and B_1' on the line A_1B_1 such that $A_1'A \perp AB$ and $B_1'B \perp AB$. Similarly, construct points C_1' and D_1' on the line C_1D_1 such that $C_1'C \perp CD$ and $D_1'D \perp CD$ (Fig. 132).

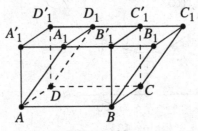

Figure 132.

The volume of the new prism $ABCDA_1'B_1'C_1'D_1'$ is again V. As one can immediately see, the surface area of the new prism is not larger than the surface area of the initial prism. Using one more construction of this type, we get a right prism with base $ABCD$ having the same volume V and surface area not larger than the surface area of the initial prism.

Next, consider an arbitrary double quadrilateral prism consisting of an "upper" prism $ABCDA_1B_1C_1D_1$ and a "lower" prism $A_2B_2C_2D_2ABCD$. Using the above argument, we may assume that both prisms are right, i.e., that the double prism is simply an ordinary right quadrilateral prism of volume V. Using again an argument similar to the above, one observes that it is enough to consider the case of a rectangular parallelepiped of volume V. Let a, b, and c be the lengths of the sides of the parallelepiped. Then $V = abc$, while for the surface area S we have

$$S = 2(ab + bc + ca) \geq 6\sqrt[3]{(ab)(bc)(ca)} = 6V^{2/3},$$

where equality holds if and only if $a = b = c$, i.e., when the parallelepiped is a cube.

1.4.19 Fix three points A, B, and C on the sphere. Clearly the volume of $ABCD$ is a maximum when the distance from D to the plane ABC is a maximum, i.e., when the orthogonal projection H of D on this plane coincides with the circumcenter of $\triangle ABC$. Moreover, the segment DH must contain the center O of the sphere. In what follows we consider only tetrahedra $ABCD$ with these properties.

Let R be the radius of the sphere. Fix D and the plane α of the base ABC of the tetrahedron. Then the intersection of α with the sphere is a circle in which $\triangle ABC$ is inscribed. Since the volume of $ABCD$ is a maximum when the area of ABC is a maximum, it follows from Problem 1.4.7 that we may assume that $\triangle ABC$ is equilateral.

The above arguments show that it is enough to consider only regular triangular pyramids inscribed in the sphere. In this case (Fig. 133) let $d = OH$ and let r be the circumradius of $\triangle ABC$.

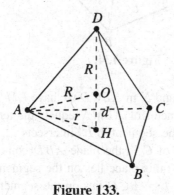

Figure 133.

Then $r = \sqrt{R^2 - d^2}$ and $[ABC] = \frac{3\sqrt{3}}{4}r^2$. For the volume V of $ABCD$ we have

$$V = \frac{1}{3}(R + d)[ABC] = \frac{\sqrt{3}}{4}(R + d)(R + d)(R - d)$$

$$= \frac{\sqrt{3}}{8}(R + d)(R + d)(2R - 2d)$$

$$\leq \frac{\sqrt{3}}{8}\left[\frac{(R + d) + (R + d) + (2R - 2d)}{3}\right]^3 = \frac{8\sqrt{3}}{27}R^3,$$

where equality holds only when $R + d = 2R - 2d$, i.e., when $R = 3d$. This is equivalent to $ABCD$ being a regular tetrahedron, i.e., all its edges have the same length.

1.4.20 Let L be a fixed point on AC. We are going to show that there exist unique points M_L in $\triangle ABD$ and N_L in $\triangle BCD$ such that the perimeter of $\triangle LM_LN_L$ is minimal among the triangles LMN with M in triangle ABD and N in triangle BCD.

Let L' and L'' be the reflections of L in the planes ABD and BCD, respectively. Denote by M_0 and N_0 the centers of the equilateral triangles ABD and BCD (Fig. 134).

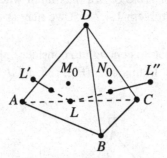

Figure 134.

For any points M in $\triangle ABD$ and N in $\triangle BCD$ we have $LM = L'M$ and $LN = L''N$, which gives that the perimeter of $\triangle LMN$ equals the length of the broken line $L'MNL''$. We claim that the segment $L'L''$ intersects $\triangle ABD$ and $\triangle BCD$. Since the orthogonal projection of C in the plane ABD coincides with M_0, the orthogonal projection L_1 of L in this plane lies on the segment AM_0. Similarly, the orthogonal projection L_2 of L in BCD lies on the segment CN_0. Let Q be the midpoint of BD. The points $L, L_1, L', L_2,$ and L'' lie in the plane AQC, and $\angle AQC < 90°$. In this plane L' is the reflection of L in the line AQ, while L'' is the reflection of L in the line CQ, so $\angle L'QL'' = 2\angle AQC < 180°$. This shows that the segment $L'L''$ intersects AQ and CQ at some points M_L and N_L, respectively (Fig. 135). It is now clear that $\triangle LM_LN_L$ has a minimum perimeter among the triangles LMN, and this perimeter is equal to $L'L''$. The latter is the length of the base of the isosceles triangle $L'L''Q$ with $L'Q = L''Q = LQ$ and

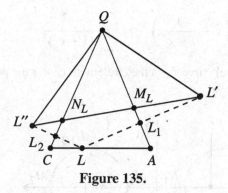

Figure 135.

$\angle L'QL'' = 2\angle AQC$. Thus $L'L''$ is minimal when LQ is shortest, i.e., when L is the midpoint of AC. In this case $M_L = M_0$ and $N_L = N_0$.

4.5 The Tangency Principle

1.5.6

(a) Let $AB = 2d$. Consider the half-planes determined by the perpendicular bi-sector of the line segment AB. We have $f(M) = MA$ for M in the half-plane containing A, and $f(M) = MB$ in the other half-plane. Hence the level curve of $f(M)$ corresponding to a number $r > 0$ is the union of two circles when $r \le d$, and the union of two arcs of circles when $r > d$ (Fig. 136).

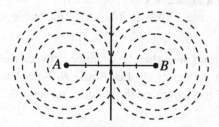

Figure 136.

(b) We may assume without loss of generality that $AB = 1$. Introduce an orthog-onal coordinate system in the plane with origin at B and such that the point A has coordinates $(1, 0)$. For a given positive number c denote by L_c the level curve of the function $f(M) = \frac{MA}{MB}$. Let $M = (x, y)$ be a point on L_c. Then $MA^2 = c^2 MB^2$ and we get $(x - 1)^2 + y^2 = c^2(x^2 + y^2)$. If $c = 1$, then L_c is the line $x = \frac{1}{2}$, i.e., the perpendicular bisector of the segment AB (Fig. 137). If $c \ne 1$, then the identity above can be written as

$$\left(x - \frac{1}{1 - c^2}\right)^2 + y^2 = \frac{c^2}{(1 - c^2)^2}.$$

Hence for $c \neq 1$ the level curve L_c is the circle with center the point $\left(\frac{1}{1-c^2}, 0\right)$ and radius $\frac{c}{|1-c^2|}$ (Fig. 137).

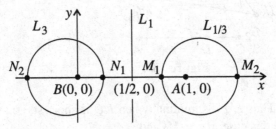

Figure 137.

The circle $L_c, c \neq 1$, is known as the circle of Apollonius for the points A and B, corresponding to the ratio c.

1.5.7 Let A, B be two fixed points such that $AB = \ell$, and let C vary along the line m parallel to AB at distance $2S/\ell$ from AB. The product of the altitudes of $\triangle ABC$ is $8S^3$ divided by the product of the three side lengths. Hence it suffices to minimize $AC \cdot BC$, which is equivalent to maximizing $\sin C$, because $AC \cdot BC = (2S)/\sin C$. Let D be the intersection of the line m and the perpendicular bisector of AB. If $\angle ADB$ is not acute, then clearly the optimal triangles are the ones with vertices C on m and with right angles at C.

Suppose that $\angle ADB$ is acute. Then it follows from Problem 1.5.1 that the optimal triangle is $\triangle ABD$.

1.5.8 Construct two parallel lines such that the distance between them is the length of the given altitude through the vertex A. Let B and B_1 be points on these lines such that BB_1 equals twice the length of the median through B. Let D be the midpoint of BB_1 (Fig. 138).

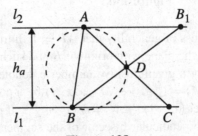

Figure 138.

Now the problem is to find a point A on ℓ_2 such that $\angle BAD$ is a maximum, which reduces to Problem 1.5.1.

1.5.9 Let C be a point on ℓ and let P and Q be the feet of the altitudes in $\triangle ABC$ through A and B. Then the points A, B, P, Q lie on a circle k with diameter AB. There are two cases to consider.

Case 1. Let k have a common point with ℓ. Then each common point of k and ℓ is a solution of the problem, since in this case $PQ = 0$.

Case 2. Let k and ℓ have no common points. Clearly either P or Q lies on a side of $\triangle ABC$. Let P lie on BC (Fig. 139). Then $\angle QBP = 90° - \angle ACB$, and the length of the chord PQ is a minimum when $\angle ACB$ is maximal, since $PQ = AB \sin \angle QBP = AB \cos \angle ACB$. Now it remains to use Problem 1.5.1.

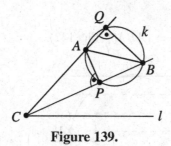

Figure 139.

1.5.10 *Hint.* Use the same argument as in the solution of Problem 1.5.1.

1.5.11 According to Problem 1.5.1 one has to construct a circle through O and A that is tangent to the given circle. There are two such circles, and their tangent points give the solutions of the problem (Fig. 140).

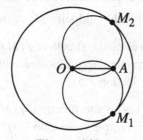

Figure 140.

1.5.12 *Answer.* The required points are the vertices of the cube that do not belong to the given diagonal.

1.5.13 The level curves of the function $f(M) = AM^2 + BM^2$ are concentric circles centered at the midpoint O of the segment AB (cf. Example 3 in Section 1.5). Hence the tangency principle implies that the minimum of $f(M)$ on l is attained at the orthogonal projection M_0 of O on l (Fig. 141).

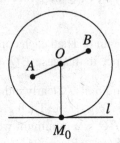

Figure 141.

1.5.14

(a) Consider the function $f(M) = [ABM]$ (Fig. 142).

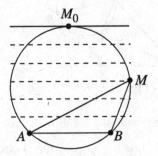

Figure 142.

The level curves of $f(M)$ are lines parallel to AB. It follows from the tangency principle that the required point is the midpoint M_0 of the larger arc $\overset{\frown}{AB}$.

(b) Use the fact that the level curves of the function $f(M) = MA^2 + MB^2$ are concentric circles whose common center coincides with the midpoint of the segment AB (cf. Example 3 in Section 1.5).

(c) Use the fact that the level curves of the function $f(M) = MA + MB$ are ellipses with foci A and B (cf. Example 7, Section 1.5).

1.5.15 The level curves of the function $f(X) = XA_1^2 + \cdots + XA_n^2$ are circles centered at the centroid G of the set of points $\{A_1, \ldots, A_n\}$ (cf. Example 5, Section 1.5). It follows from the tangency principle that $X \in M$ has to be chosen in such a way that GX is minimal. One is now left to deal with the problem described in the remark after Problem 1.5.2.

1.5.16 The solution is similar to the solution of the previous problem.

1.5.17 See the solution of Problem 1.5.3.

1.5.18 Let $\triangle ABC$ be isosceles and right-angled with $\angle C = 90°$. Introduce a coordinate system with origin C and coordinate axes CA and CB (Fig. 143).

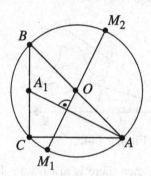

Figure 143.

Let $A = (a, 0)$, $B = (0, a)$. The level curve L_r of the function $f(M) = MA^2 + 2MB^2 - 3MC^2$ is the line $x + 2y = \frac{3a^2 - r}{2a}$ (see the Theorem, after Example 4 in Section 1.5). Let A_1 be the midpoint of BC. Then the line AA_1 is the level curve of $f(M)$ corresponding to $r = a^2$. It follows from the tangency principle that the points M_1 and M_2 where $f(M)$ achieves its minimum and maximum, respectively, are tangent points of the circumcircle of $\triangle ABC$ with lines parallel to AA_1. Clearly M_1 and M_2 are the intersection points of the circumcircle with the line through its center O and perpendicular to AA_1 (Fig. 143).

In the case of an equilateral triangle ABC use the same argument as above.

1.5.19 According to the tangency principle the maximum (minimum) of the function $f(M) = \frac{AM}{BM}$ is attained at points where a level curve L_c of $f(M)$ is tangent to the line l. So, we may assume that $c \neq 1$. Set $AB = m$ and let d be the distance between the parallel lines AB and l. From the solution of Problem 1.5.6 (b) we know that for any $c > 0, c \neq 1$, the level curve L_c of $f(M)$ is a circle with center on the line AB and radius $\frac{mc}{|1-c^2|}$. Such a circle is tangent to the line l if its radius is equal to d, i.e., when $|1 - c^2| = \frac{m}{d}c$. Solving this equation for c, we conclude that the maximum and the minimum of $f(M)$ on l are given respectively by

$$\frac{1}{2}\left(\frac{m}{d} + \sqrt{\left(\frac{m}{d}\right)^2 + 4}\right) \text{ and } \frac{1}{2}\left(-\frac{m}{d} + \sqrt{\left(\frac{m}{d}\right)^2 + 4}\right).$$

1.5.20 The level curves of the function

$$f(X) = d(X, \ell_1) + d(X, \ell_2),$$

where X is a point in the interior of the angle, are line segments perpendicular to the bisector of the angle (see Example 6, Section 1.5). Thus the required points X can be found in the following way: Move a line through the vertex O keeping it perpendicular to the angle bisector until it meets a point (points) of M. The point(s) obtained in this way give the solution (Fig. 144).

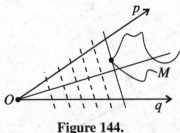

Figure 144.

Notice that if M is a polygon, then there is always a solution of the problem that is a vertex of M (Figs. 145, 146).

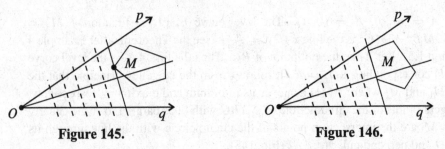

Figure 145.   **Figure 146.**

In the case of a circle, the solution is given by the tangent point of a tangent line to the circle perpendicular to the angle bisector (Fig. 147).

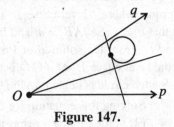

Figure 147.

1.5.21 *Hint.* Show that the level curves of the functions

$$f(\ell) = OC + OD - CD, \quad g(\ell) = OC + OD + CD,$$

depending on a variable line ℓ, consist of the tangent lines to the larger and the smaller arcs, respectively, of the circles inscribed in the angle (Figs. 148, 149). Then use the tangency principle.

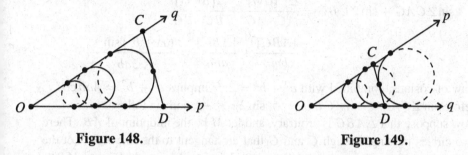

Figure 148.   **Figure 149.**

1.5.22 Let ABM be one of the given triangles, where AB is the given side. Then $MA + MB = 2p - AB$, and as we know from Example 7, Section 1.5, the locus of the points M with this property is an ellipse with foci A and B (Fig. 150).

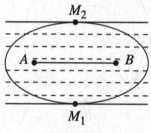

Figure 150.

The level curves of the function $f(M) = [ABM]$ are lines parallel to the axis AB of the ellipse. Now the tangency principle implies that the solution of the problem is given by the isosceles triangle having the required properties.

1.5.23 The required maximum is equal to $2/\sqrt{3}$. We first prove that

$$(1) \qquad \sin \angle CAG + \sin \angle CBG \leq \frac{2}{\sqrt{3}}$$

if the circumcircle of triangle ACG is tangent to the line AB, and then handle the case of an arbitrary triangle ABC. So, let the circumcircle of $\triangle ACG$ be tangent to AB. We use the standard notation for the elements of $\triangle ABC$. By the power-of-a-point theorem and the well-known median formula (see Glossary) we have

$$\frac{c^2}{4} = MA^2 = MG \cdot MC = \frac{1}{3}m_c^2 = \frac{1}{12}(2a^2 + 2b^2 - c^2),$$

yielding $a^2 + b^2 = 2c^2$. Using the median formula again gives $m_a = \frac{\sqrt{3}}{2}b$, $m_b = \frac{\sqrt{3}}{2}a$. Then

$$\sin \angle CAG + \sin \angle CBG = \frac{2[ACG]}{AC \cdot AG} + \frac{2[BCG]}{BC \cdot BG}$$

$$= \frac{[ABC]}{bm_a} + \frac{[ABC]}{am_b} = \frac{(a^2 + b^2) \sin \gamma}{\sqrt{3}ab}.$$

The law of cosines, combined with $a^2 + b^2 = 2c^2$, implies $a^2 + b^2 = 4ab \cos \gamma$. Therefore $\sin \angle CAG + \sin \angle CBG = \frac{2}{\sqrt{3}} \sin 2\gamma \le \frac{2}{\sqrt{3}}$, and (1) follows.

Now suppose that $\triangle ABC$ is arbitrary, and let M be the midpoint of AB. There are two circles passing through C and G that are tangent to the line AB. Let the corresponding points of tangency be A_1 and B_1, lying on the rays $MA^{\rightarrow}$ and $MB^{\rightarrow}$, respectively (Fig. 151).

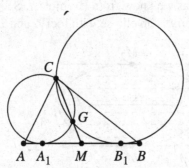

Figure 151.

Since $MA_1^2 = MG \cdot MC = MB_1^2$ by the power-of-a-point theorem and $CG : GM = 2 : 1$, G is the centroid of $\triangle A_1 B_1 C$ as well. Moreover, A and B are exterior to the two circles unless $A = A_1$ and $B = B_1$. It is straightforward now that $\angle CAG \le \angle CA_1G$, $\angle CBG \le \angle CB_1G$. Thus, assuming $\angle CA_1G$ and $\angle CB_1G$ acute we conclude by the special case already settled that

$$\sin \angle CAG + \sin \angle CBG \le \sin \angle CA_1G + \sin \angle CB_1G \le \frac{2}{\sqrt{3}}.$$

Thus we are left with the proof of (1) in the case that one of $\angle CA_1G$ and $\angle CB_1G$ is right or obtuse.

Let, for instance, $\angle CA_1G \ge 90°$; then $\angle CB_1G$ is acute. Denote by a_1, b_1, c_1 the side lengths of $\triangle A_1 B_1 C$ and let $\gamma_1 = \angle A_1 C B_1$. We obtain from $\triangle CA_1G$ that $CG^2 > CA_1^2 + A_1G^2$, that is,

$$\frac{1}{9}(2a_1^2 + 2b_1^2 - c_1^2) > b_1^2 + \frac{1}{9}(2b_1^2 + 2c_1^2 - a_1^2).$$

We have $a_1^2 + b_1^2 = 2c_1^2$, and the above inequality takes the form $a_1^2 > 7b_1^2$. Now set $x = b_1^2/a_1^2$. The argument in the proof of the special case also gives

$$\sin \angle CB_1G = \frac{2[B_1CG]}{B_1C \cdot B_1G} = \frac{b_1 \sin \gamma_1}{a_1\sqrt{3}}$$

$$= \frac{b_1}{a_1\sqrt{3}}\sqrt{1 - \left(\frac{a_1^2 + b_1^2}{4a_1 b_1}\right)^2} = \frac{1}{4\sqrt{3}}\sqrt{14x - x^2 - 1} = f(x).$$

Since $x < 1/7$, it follows that $f(x) < f(1/7) = 1/7$. Therefore

$$\sin \angle CAG + \sin \angle CBG < 1 + \sin \angle CB_1G < 1 + \frac{1}{7} < \frac{2}{\sqrt{3}}.$$

4.6 Isoperimetric Problems

2.1.5 The solution follows from Heron's formula for the area F of a triangle with sides a, b, c, which can be written as

$$F^2 = s(s - c)[c^2 - (a - b)^2],$$

where s is the semiperimeter of the triangle.

2.1.6 This follows immediately from the previous problem.

2.1.7 The area of a parallelogram with sides a and b and angle α between them is given by $S = ab \sin \alpha$. Hence $S \leq ab \leq \left(\frac{a+b}{2}\right)^2$, where equality holds if $a = b$ and $\alpha = 90°$, i.e., when the parallelogram is a square.

2.1.8 *Hint.* Use Problem 2.1.6.

2.1.9 It is easily seen that we may consider only convex quadrilaterals of area 1. Let $ABCD$ be such a quadrilateral and AB its longest side. Denote by D' and C' the reflections of D and C in the line AB (Fig. 152).

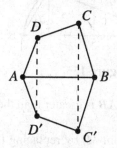

Figure 152.

Then the area of the hexagon $AD'C'BCD$ is equal to 2, and by the isoperimetric theorem for hexagons it follows that

$$BC + CD + DA = \frac{1}{2}(AD' + D'C' + C'B + BC + CD + DA) \geq 2\sqrt[4]{3}.$$

Hence the minimum of the sum $BC + CD + DA$ is attained only for trapezoids $ABCD$ such that $BC = CD = DA = \frac{2}{3}\sqrt[4]{3}$ and $AB = \frac{4}{3}\sqrt[4]{3}$.

2.1.10 *Hint.* Use the idea of the solution of the previous problem and Problem 2.1.3.

2.1.11 Denote by $a_1, a_2, \ldots, a_n$ the successive sides of M, so that a_1 is the shortest and a_p (for some $p > 1$) the longest. We now construct a new n-gon M' as follows. We leave the sides $a_p, a_{p+1}, \ldots, a_n$ unchanged. Then starting at the "free" end of a_p we construct consecutively chords of lengths $a_1, a_{p-1}, \ldots, a_2$ (Figs. 153, 154).

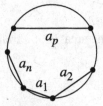

Figure 153.

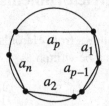

Figure 154.

The resulting n-gon M' has the same sides and the same area as M, and its shortest and longest sides are next to each other. Moreover, we have $a_1 \leq s \leq a_p$, where each equality holds only when M is a regular n-gon. Thus, we may assume that $a_1 < s < a_n$. Consider the arc L determined by the chords a_1 and a_p. Using the notation from Fig. 155, where C' and D are points on L such that $AC' = BC$ and $BD = s$, we have that D is on the arc $\overset{\frown}{C'C}$.

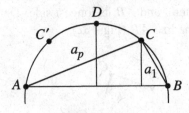

Figure 155.

Therefore the distance from D to AB is greater than the distance from C to AB, i.e., $[ABC] < [ABD]$.

Let M'' be the n-gon obtained from M' by replacing the vertex C by D. Then M'' has the desired property.

2.1.12 To solve the problem one has to repeat the construction used in the solution of the previous problem at most $n - 1$ times.

2.1.13 Let $ABCD$ be a quadrilateral with vertices on the given four circles and let O be the center of the square. Suppose that the quadrilateral $ABCD$ is not convex, say $\angle ABC > 180°$. Then the point B and the center O_b of the circle containing B lie on different sides of the line AC. Denote by B' the intersection point of the perpendicular to AC through B with the circle with center O_b. Then $AB' > AB$, $CB' > CB$, and therefore the perimeter of $AB'CD$ is not less than the perimeter of $ABCD$. Hence we may assume that $ABCD$ is a convex quadrilateral.

Let k be the circle with center O such that the given four circles are internally tangent to it. Denote by A_1, B_1, C_1, and D_1 the intersection points of k with the rays OA, OB, OC, and OD, respectively. Since quadrilateral $ABCD$ is convex and lies in $A_1 B_1 C_1 D_1$, it follows that its perimeter is not larger than that of $A_1 B_1 C_1 D_1$. On the other hand, it follows from Problem 2.1.12 that the perimeter of $A_1 B_1 C_1 D_1$ is not larger than the perimeter of a square inscribed in k. Hence the desired quadrilateral has vertices at the tangent points of k with the given four circles (Fig. 156).

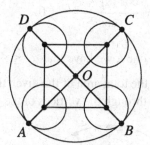

Figure 156.

2.1.14 Set $MA_k = x_k$, $A_k A_{k+1} = a_k$ and $\angle MA_k A_{k+1} = \alpha_k$ for $k = 1, 2, \ldots, n$ ($A_{n+1} = A_1$). Let S be the area of $A_1 A_2 \ldots A_n$. Then

$$2S = \sum_{k=1}^{n} a_k x_k \sin \alpha_k.$$

By the law of cosines for $\triangle M A_k A_{k+1}$ we get

$$x_{k+1}^2 = x_k^2 + a_k^2 - 2x_k a_k \cos \alpha_k.$$

Summing up these equalities for $k = 1, 2, \ldots, n$ gives

$$\sum_{k=1}^{n} a_k^2 = 2 \sum_{k=1}^{n} a_k x_k \cos \alpha_k.$$

On the other hand, the root mean square–arithmetic mean inequality together with the isoperimetric theorem for n-gons gives

$$\sum_{k=1}^{n} a_k^2 \geq \frac{1}{n}\left(\sum_{k=1}^{n} a_k\right)^2 \geq 4S \tan \frac{\pi}{n}.$$

Hence

$$\sum_{k=1}^{n} a_k x_k \cos \alpha_k \geq \sum_{k=1}^{n} a_k x_k \tan \frac{\pi}{n} \sin \alpha_k,$$

which can be written as

$$\sum_{k=1}^{n} a_k x_k \frac{\cos \left(\alpha_k + \frac{\pi}{n}\right)}{\cos \frac{\pi}{n}} \geq 0.$$

Suppose that $\alpha_k > \frac{\pi(n-2)}{2n}$ for $k = 1, 2, \ldots, n$. Then $\frac{3\pi}{2} > \alpha_k + \frac{\pi}{n} > \frac{\pi}{2}$ and therefore $\cos\left(\alpha_k + \frac{\pi}{n}\right) < 0$ for $k = 1, 2, \ldots, n$. Thus

$$\sum_{k=1}^{n} a_k x_k \frac{\cos \left(\alpha_k + \frac{\pi}{n}\right)}{\cos \frac{\pi}{n}} < 0,$$

a contradiction. Hence for at least one k we have that $\alpha_k \leq \frac{\pi(n-2)}{2n}$.

2.1.15 Assume the contrary. Then the total area of the given three triangles is equal to 3. Consider the ends of the radii through the vertices of these triangles (Fig. 157).

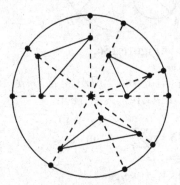

Figure 157.

They form a polygon with at most 9 vertices. By Problem 2.1.12 it follows that its area is not larger than the area of a regular 9-gon inscribed in the unit circle. Hence the total area of the three triangles is less than the area of a regular 12-gon inscribed in the unit circle, which is just 3, a contradiction.

Note that the solution also follows by the fact that if a triangle of area 1 lies in a unit circle with center O, then O lies in its interior or on its boundary. We leave this as an exercise to the reader.

2.1.16 *Hint.* First show that it is enough to consider convex n-gons. Then, using appropriate symmetries, show that the shortest and longest sides of the n-gon can be assumed next to each other. One can then use the method from the solution of Problem 2.1.11.

2.1.17 Consider the position of the rope for which it forms an arc of a circle, while the stick is the corresponding chord in the circle (Fig. 158).

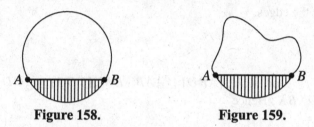

Figure 158.   **Figure 159.**

Add to the sector of the disk bounded by the rope and the stick the remaining sector of the disk (the one marked in Fig. 159). It now follows from the isoperimetric theorem that in this position the rope and the stick bound a region of maximum possible area (Fig. 158).

2.1.18 Consider an arbitrary figure cut off from the given angle. Using $2n - 1$ consecutive symmetries with respect to lines, one gets a region in the plane bounded by a closed curve of length $2n\ell$ (Fig. 160).

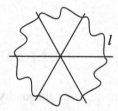

Figure 160.

The isoperimetric theorem now yields that the initial curve must be an arc of a circle with center at the vertex of the angle.

2.1.19 Let α be the angle between the planes of the base and a lateral face. It is easy to see that $V = \dfrac{S^{3/2}}{\sqrt[3]{n \tan\left(\frac{180°}{n}\right)}} f(\alpha)$, where $f(\alpha) = \dfrac{\sqrt{\cos\alpha}(1-\cos\alpha)}{1+\cos\alpha}$. Set $t = \cos\alpha$.

We have to find the maximum of the function $g(t) = \dfrac{\sqrt{t}(1-t)}{1+t}$ for $t \in (0, 1)$. Since

$g'(t) = \frac{1-3t}{2(1+t^2)\sqrt{t(1-t)}}$, it is easy to observe that $g(t)$ achieves its maximum on the interval $(0, 1)$ at $t = \frac{1}{3}$, i.e., the maximum value of $f(\alpha)$ is achieved when $\cos \alpha = \frac{1}{3}$. The maximal volume is equal to $\frac{\sqrt{2}}{12} \cdot \frac{S^{3/2}}{\sqrt{n \tan\left(\frac{180°}{n}\right)}}$.

2.1.20 Of all parallelograms with sides of given lengths, the rectangle has largest area. It is also clear that of all parallelepipeds with edges of given lengths the right rectangular parallelepiped has maximal volume. Then the arithmetic mean–geometric mean inequality implies $V^3 = (abc)^3 \leq \left(\frac{a+b+c}{3}\right)^3$, where equality holds when $a = b = c$. Hence the cube has maximum volume among the parallelepipeds with a given sum of the edges.

2.1.21

(a) Clearly $[ABC] \leq \frac{1}{2}AB \cdot CM$, $[ABD] \leq \frac{1}{2}AB \cdot DM$, $[CDA] \leq \frac{1}{2}CD \cdot AK$, $[CDB] \leq \frac{1}{2}CD \cdot BK$. Hence

$$S = [ABC] + [ABD] + [CDA] + [CDB]$$

$$\leq \frac{1}{2}AB\,(CM + DM) + \frac{1}{2}CD\,(AK + BK).$$

Since MK is a median in $\triangle AKB$ and $\triangle CMD$, one gets

$$4MK^2 = 2(AK^2 + BK^2) - AB^2 = 2(CM^2 + DM^2) - CD^2.$$

This implies

$$AK^2 + BK^2 = \frac{4c^2 + a^2}{2}, \quad CM^2 + DM^2 = \frac{4c^2 + b^2}{2}.$$

Now the root mean square–arithmetic mean inequality gives

$$\frac{CM + DM}{2} \leq \sqrt{\frac{CM^2 + DM^2}{2}} = \frac{1}{2}\sqrt{4c^2 + b^2},$$

$$\frac{AK + BK}{2} \leq \sqrt{\frac{AK^2 + BK^2}{2}} = \frac{1}{2}\sqrt{4c^2 + a^2}.$$

Using these yields $S \leq \frac{1}{2}(a\sqrt{4c^2 + b^2} + b\sqrt{4c^2 + a^2})$. Equality holds if $AB \perp MK$, $CD \perp MK$, and $AB \perp CD$, in which case S is a maximum.

(b) For the volume V of $ABCD$ we have $V \leq \frac{1}{3}[ABK] \cdot CD$. On the other hand, $[ABK] \leq \frac{1}{2}AB \cdot MK$. These inequalities imply $V \leq \frac{abc}{6}$, where equality holds when $AB \perp CD$, $AB \perp MK$, and $CD \perp MK$. In this case the volume V of $ABCD$ is a maximum.

2.1.22 Let α be the plane through B that is perpendicular to AB. The projection of $\triangle ACD$ onto α is $\triangle BEF$ (Fig. 161).

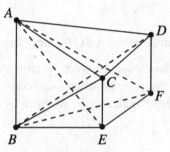

Figure 161.

Then the volume V of the tetrahedron $ABCD$ is equal to $\frac{1}{3}AB \cdot [BEF]$. This follows from the fact that the volumes of tetrahedra $ABCD$ and $ABEF$ are equal to the volume of the tetrahedron $ABCF$. Hence V is a maximum when the area of $\triangle BEF$ is a maximum. Since of all triangles with a given perimeter the equilateral triangle has a maximum area (Problem 1.2.1), it is enough to find out when the perimeter of $\triangle BEF$ is a maximum. To do so, "unfold" the planes $FDCE$ and CEB onto the plane $ABFD$ (Fig. 162).

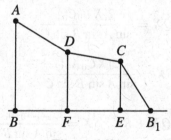

Figure 162.

Then the perimeter of $\triangle BEF$ equals BB_1. The latter is a maximum when the segments AD, CD, and CB_1 form the same angle γ with the side AB such that

$$\cos \gamma = \frac{AB}{AD + DC + CB}.$$

In this case the perimeter of $\triangle BEF$ is a constant and therefore it has a maximum area when $BE = BF = EF$. Hence the tetrahedron $ABCD$ has a maximum volume when $AD = CD = CB$, and these three edges make equal angles with AB.

2.1.23 It follows from the previous problem that when $AB = h$ is fixed, the maximum volume of the tetrahedron $ABCD$ is equal to

$$V = \frac{p}{9\sqrt{3}} h(p - h),$$

where p is the perimeter of the quadrilateral $ABCD$. Since $h(p-h) \le \left(\frac{h+p-h}{2}\right)^2 = \frac{p^2}{4}$, it follows that $V \le \frac{p^3}{36\sqrt{3}}$, where equality holds when $h = \frac{p}{2}$. In this case the skew quadrilateral $ABCD$ has sides of equal lengths and equal angles between any two adjoining sides. Denoting these angles by γ, we have by Problem 2.1.22 that $\cos \gamma = \frac{1}{3}$.

4.7 Extremal Points in Triangle and Tetrahedron

2.2.5 The statement follows immediately from Problem 1.1.3.

2.2.6 Let A_0, B_0, and C_0 be the feet of the perpendiculars from X to BC, CA, and AB, respectively. Then

$$MX \cdot NX = \frac{XB_0}{\sin A} \cdot \frac{XA_0}{\sin B} = \frac{2[XA_0B_0]}{\sin A \sin B \sin C}.$$

Similarly

$$PX \cdot QX = \frac{2[XB_0C_0]}{\sin A \sin B \sin C}$$

and

$$RX \cdot SX = \frac{2[XC_0A_0]}{\sin A \sin B \sin C}.$$

Hence

$$MX \cdot NX + PX \cdot QX + RX \cdot SX = \frac{2[A_0B_0C_0]}{\sin A \sin B \sin C}.$$

Now using Problem 2.2.2 we conclude that the given sum is a maximum when X is the circumcenter of $\triangle ABC$.

2.2.7

(a) Let AA_1 and MM_1 be the altitudes of triangles ABC and MBC, respectively. Then

$$[AMB] + [AMC] = [ABC] - [BMC] = \frac{(AA_1 - MM_1)BC}{2} \le \frac{AM \cdot BC}{2}$$

with equality only if $AM \perp BC$. Similarly,

$$[AMB] + [BMC] \le \frac{BM \cdot AC}{2}$$

and

$$[BMC] + [AMC] \le \frac{CM \cdot AB}{2}.$$

Adding the above inequalities gives

$$AM \cdot BC + BM \cdot AC + CM \cdot AB \ge 4[ABC].$$

Thus, the minimum of the given sum is $4[ABC]$, and it is attained only if M is the orthocenter of $\triangle ABC$.

(b) Let E and F be points such that $BCME$ and $BCAF$ are both parallelograms (Fig. 163). Then $EMAF$ is also a parallelogram.

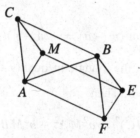

Figure 163.

Hence

$$AF = EM = BC, \quad EF = AM, \quad EB = CM, \quad BF = AC.$$

Applying Ptolemy's inequality (Problem 3.2.6) to quadrilaterals $ABEF$ and $AEBM$, we have

$$AB \cdot AM + BC \cdot CM = AB \cdot EF + AF \cdot BE \ge AE \cdot BF = AE \cdot AC,$$

$$BM \cdot AE + AM \cdot CM = BM \cdot AE + AM \cdot BE \ge AB \cdot EM = AB \cdot BC.$$

Therefore

$$MA \cdot MB \cdot AB + MB \cdot MC \cdot BC + MC \cdot MA \cdot CA$$

$$= MB(MA \cdot AB + MC \cdot BC) + MC \cdot MA \cdot CA$$

$$\ge MB \cdot AE \cdot AC + MC \cdot MA \cdot CA$$

$$= AC(MB \cdot AE + MC \cdot MA) \ge AC \cdot AB \cdot BC.$$

Equality holds if and only if both $ABEF$ and $AEBM$ are cyclic, which implies that $AFEM$ is cyclic. Since $AFEM$ is a parallelogram it follows that $AM \perp EM$, i.e., $AM \perp BC$. Since $AEBM$ is cyclic, $\angle ABE = \angle AME$, which implies $BE \perp AB$, i.e., $CM \perp AB$. Thus M is the orthocenter of $\triangle ABC$.

Remark. The inequality

$$MA \cdot MB \cdot AB + MB \cdot MC \cdot BC + MC \cdot MA \cdot CA \geq AB \cdot BC \cdot CA$$

can be proved also by using complex numbers. Indeed, let M be the origin of the complex plane and let the complex coordinates of A, B, C be u, v, w, respectively. Then the given inequality can be written as

$$|uv(u - v)| + |vw(v - w)| + |wu(w - u)| \geq |(u - v)(v - w)(w - u)|.$$

But it is easily checked that

$$uv(u - v) + vw(v - w) + wu(w - u) = -(u - v)(v - w)(w - u),$$

and the inequality above follows by the triangle inequality.

2.2.8 Set $AB = c, BC = a, CA = b$. Then

$$0 \leq (a\overrightarrow{MA} + b\overrightarrow{MB} + c\overrightarrow{MC})^2 = a^2 MA^2 + b^2 MB^2 + c^2 MC^2 +$$
$$+ 2ab(\overrightarrow{MA}, \overrightarrow{MB}) + 2bc(\overrightarrow{MB}, \overrightarrow{MC}) + 2ca(\overrightarrow{MC}, \overrightarrow{MA}).$$

From the law of cosines it follows that

$$2(\overrightarrow{MA}, \overrightarrow{MB}) = MA^2 + MB^2 - c^2,$$
$$2(\overrightarrow{MB}, \overrightarrow{MC}) = MB^2 + MC^2 - a^2,$$
$$2(\overrightarrow{MC}, \overrightarrow{MA}) = MC^2 + MA^2 - b^2.$$

Plugging these in the above inequality gives

$$(a^2 + ab + ac)MA^2 + (b^2 + ba + bc)MB^2 + (c^2 + ca + cb)MC^2$$
$$-abc^2 - bca^2 - cab^2 \geq 0,$$

which is equivalent to

$$aMA^2 + bMB^2 + cMC^2 \geq abc.$$

Equality occurs if and only if

(1) $$a\overrightarrow{MA} + b\overrightarrow{MB} + c\overrightarrow{MC} = 0.$$

So, we have to find the points M satisfying (1). Note that the lines AM and BC are not parallel, since otherwise the vectors $\overrightarrow{MA}$ and $b\overrightarrow{MB} + c\overrightarrow{MC}$ are not collinear. Denote by A_1 the intersection point of AM and BC. Then

$$0 = a\overrightarrow{MA} + b(\overrightarrow{MA_1} + \overrightarrow{A_1B}) + c(\overrightarrow{MA_1} + \overrightarrow{A_1C})$$
$$= (a\overrightarrow{MA} + (b+c)\overrightarrow{MA_1}) + (b\overrightarrow{A_1B} + c\overrightarrow{A_1C}).$$

The first vector on the right-hand side is collinear to $\overrightarrow{AM}$, whereas the second one is collinear to $\overrightarrow{BC}$. Hence each of them is 0. This implies

$$\frac{A_1B}{A_1C} = \frac{c}{b} = \frac{AB}{AC},$$

i.e., AA_1 is the angle bisector of $\angle BAC$.

Applying the same reasoning to BM and CM, we conclude that the only point M satisfying (1) is the incenter of triangle ABC.

Remark. Using the same reasoning as above one can solve the following more general problem: *Given a triangle ABC and real numbers p, q, r such that $p + q + r > 0$, find the points M in the plane such that*

$$pMA^2 + qMB^2 + rMC^2$$

is a minimum.

Note that the desired minimum is equal to $\frac{qra^2 + prb^2 + pqc^2}{p+q+r}$ and it is attained at the point M such that

$$\overrightarrow{AM} = \frac{q}{p+q+r}\overrightarrow{AB} + \frac{r}{p+q+r}\overrightarrow{AC}.$$

2.2.9 Let α, β, γ be the angles A, B, C, respectively. Set $\alpha_1 = \angle MAB$ and $\alpha_2 = \angle MAC$. We have

$$\frac{MB' \cdot MC'}{MA^2} = \sin\alpha_1 \sin\alpha_2.$$

Observe that

$$\sin\alpha_1 \sin\alpha_2 = \frac{1}{2}(\cos(\alpha_1 - \alpha_2) - \cos(\alpha_1 + \alpha_2)) \le \frac{1}{2}(1 - \cos\alpha) = \sin^2\frac{\alpha}{2}.$$

Hence

$$\frac{MB' \cdot MC'}{MA^2} \le \sin^2 \frac{\alpha}{2}.$$

Likewise,

$$\frac{MA' \cdot MC'}{MB^2} \le \sin^2 \frac{\beta}{2} \quad \text{and} \quad \frac{MB' \cdot MA'}{MC^2} \le \sin^2 \frac{\gamma}{2}.$$

Therefore

$$\frac{MA' \cdot MB' \cdot MC'}{MA \cdot MB \cdot MC} \le \sin \frac{\alpha}{2} \sin \frac{\beta}{2} \sin \frac{\gamma}{2}$$

with equality if and only if M is the incenter of triangle ABC.

2.2.10 One has to consider two cases.

Case 1. All angles of $\triangle ABC$ are less than or equal to $90°$. We will show that $m(X)$ is maximal when X coincides with the circumcenter O of $\triangle ABC$. Let O_1, O_2, and O_3 be the midpoints of the sides BC, AC, and AB, respectively. For any point $X \ne O$ in the quadrilateral AO_3OO_2 we have $m(X) = AX < AO = R = m(O)$ (Fig. 164).

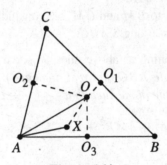

Figure 164.

In the same way we see that $m(X) < m(O)$ when $X \ne O$ lies in the quadrilateral BO_1OO_3 or CO_2OO_1.

Case 2. Triangle ABC has an obtuse angle. Assume, for example, that $\gamma > 90°$. We may also assume that $\alpha \le \beta$. Denote by D and E the midpoints of the sides BC and CA, and by F and G the intersection points of the perpendicular bisectors of the sides BC and CA with AB (Fig. 165).

Then $AG = GC = x$ and $BF = CF = y$. Since $x = \frac{b}{2\cos\alpha}$, $y = \frac{a}{2\cos\beta}$, the law of sines for $\triangle ABC$ gives

$$\frac{x}{y} = \frac{b\cos\beta}{a\cos\alpha} = \frac{\sin\beta}{\sin\alpha} \cdot \frac{\cos\beta}{\cos\alpha} = \frac{\sin 2\beta}{\sin 2\alpha} \ge 1,$$

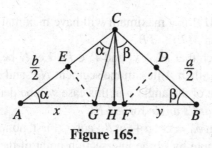

Figure 165.

where equality holds only when $\alpha = \beta$. Next, $\alpha + \beta < 90° < \gamma$ implies $\angle ACF = \gamma - \beta > \alpha$, so in $\triangle AFC$ we have $AF > FC = y$.

Let H be the foot of the altitude through C. If X is in $\triangle AHC$, then it lies either inside the circle with diameter CG (the circumcircle of the quadrilateral $CEGH$) or inside the circle with diameter AG. In both cases $m(X) \leq x = AG = CG$, where equality holds when $X = G$. Similarly, if X lies in $\triangle BCH$ we have $m(X) \leq y$ with equality only when $X = F$.

Hence, if $\alpha < \beta$, then $x > y$, and $m(X)$ is a maximum precisely when $X = G$. If $\alpha = \beta$, $m(X)$ attains its maximum when $X = G$ or $X = F$.

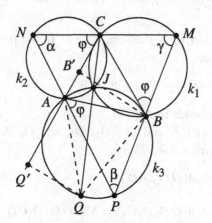

Figure 166.

2.2.11 We use the notation in Fig. 166. We have $\angle BCM = 180° - \varphi - \gamma$, so $\angle BMC = \gamma$. Similarly, $\angle CNA = \alpha$ and $\angle APB = \beta$. This means that the point M lies on an arc of a circle k_1 (the locus of the points X such that $\angle BXC = \gamma$), N on an arc of a circle k_2, and P on an arc of a circle k_3. It is easy to see that k_1, k_2, and k_3 intersect at point J; this is the so-called *Brokard's point* for $\triangle ABC$. Since all triangles MNP satisfying the assumptions of the problem are similar to

$\triangle CAB$, the one for which MP is a maximum will have maximal area. It is clear that MP is a maximum when $MP \perp JB$.

Construct $MP \perp BJ$ such that $M \in k_1$ and $P \in k_3$. Let N be the intersection point of the line MC with k_2; then A lies on the segment NP and $\triangle MNP$ has the desired properties. The value of the angle φ in this case will be denoted by φ_0, and $\omega = 90° - \varphi_0$ is called *Brokard's angle* for $\triangle ABC$.

We will now show that $\tan \varphi_0 = \cot \alpha + \cot \beta + \cot \gamma$. First, notice that $\angle JAB = \angle JBC = \angle JCA = \omega$. Denote by Q the intersection point of the line CJ and k_3 (Fig. 166). Then

$$\angle QBA = \angle QJA = \angle ACJ + \angle JAC = \omega + (\alpha - \omega) = \alpha.$$

In particular, $BQ \parallel AC$.

In a similar way one obtains $\angle QAB = \beta$. Let Q' and B' be the projections of Q and B, respectively, on the line AC. Then $QQ' = BB'$ and

$$\tan \varphi_0 = \cot \omega = \frac{CQ'}{QQ'} = \frac{CB'}{BB'} + \frac{AB'}{BB'} + \frac{AQ'}{QQ'} = \cot \gamma + \cot \alpha + \cot \beta.$$

This equality determines the angle φ_0 uniquely.

Remark. There is another Brokard's point J', which is determined by $\angle J'AC = \angle J'CB = \angle J'BA$.

2.2.12 We will use the notation from Problem 2.2.13. In the present case $S_1 = S_2 = S_3 = S_4 = S$. Notice that

$$x_1 + x_2 + x_3 + x_4 = h,$$

where h is the length of the altitude in $ABCD$.

Let O be the center of $ABCD$ and O_1, O_2, O_3, and O_4 the centers of the corresponding faces of the tetrahedron. Then

$$\text{Vol}(OO_1O_2O_3) = \text{Vol}(OO_2O_3O_4)$$

$$= \text{Vol}(OO_1O_3O_4) = \text{Vol}(OO_1O_2O_4) = \frac{1}{4}V,$$

where $V = \text{Vol}(O_1O_2O_3O_4)$. For any point X in $ABCD$ we have

$$\frac{\text{Vol}(XX_1X_2X_3)}{\frac{1}{4}V} = \frac{\text{Vol}(XX_1X_2X_3)}{\text{Vol}(OO_1O_2O_3)} = \frac{x_1x_2x_3}{r \cdot r \cdot r},$$

where $r = \frac{h}{4} = OO_1 = OO_2 = OO_3 = OO_4$. Hence

$$\text{Vol}(XX_1X_2X_3) = \frac{16V}{h^3} x_1x_2x_3.$$

One obtains similar expressions for $\mathrm{Vol}(XX_2X_3X_4)$, $\mathrm{Vol}(XX_1X_3X_4)$, and $\mathrm{Vol}(XX_1X_2X_4)$. Summing these, it follows that

$$\mathrm{Vol}(X_1X_2X_3X_4) = \frac{16V}{h^3}[x_1x_2x_3 + x_2x_3x_4 + x_1x_3x_4 + x_1x_2x_4].$$

Set $a = x_1 + x_2$ and $b = x_3 + x_4$. Then $a + b = h$, and therefore

$\mathrm{Vol}(X_1X_2X_3X_4)$

$$= \frac{16V}{h^3}[x_1x_2b + ax_3x_4] \le \frac{16V}{h^3}\left[\left(\frac{x_1 + x_2}{2}\right)^2 b + a\left(\frac{x_3 + x_4}{2}\right)^2\right]$$

$$= \frac{16V}{h^3} \cdot \frac{a^2b + ab^2}{4} = \frac{4V}{h^2}ab \le \frac{4V}{h^2}\left(\frac{a+b}{2}\right)^2 = V.$$

Equality holds when $x_1 = x_2$, $x_3 = x_4$, and $a = b$, i.e., when $X = O$. Thus the required point X is the center of $ABCD$.

2.2.13

(a) Let X lie on CD and $\frac{x_3}{S_3} = \frac{x_4}{S_4}$. Then

$$\frac{\mathrm{Vol}(ABDX)}{\mathrm{Vol}(ABCX)} = \frac{x_3 S_3}{x_4 S_4} = \frac{S_3^2}{S_4^2}.$$

Let u and v be the distances from D and C, respectively, to the plane ABX. Then $\frac{DX}{XC} = \frac{u}{v}$ (Fig. 167).

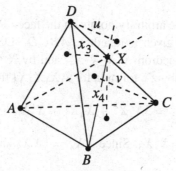

Figure 167.

On the other hand,

$$\frac{S_3^2}{S_4^2} = \frac{V_{ABDX}}{V_{ABCX}} = \frac{u \cdot [ABX]}{v \cdot [ABX]} = \frac{u}{v},$$

so $\frac{DX}{XC} = \frac{S_3^2}{S_4^2}$. Clearly there exists a unique point M on CD with $\frac{DM}{MC} = \frac{S_3^2}{S_4^2}$. It follows from the above arguments that if X lies in the plane ABM, then $\frac{x_3}{S_3} = \frac{x_4}{S_4}$, and conversely, if the latter is true, then X lies in the plane ABM. In the same way one constructs points $N \in AD$ and $P \in BD$ such that the set of points X with $\frac{x_1}{S_1} = \frac{x_4}{S_4}$ coincides with the plane BCN, while the set of points X with $\frac{x_2}{S_2} = \frac{x_4}{S_4}$ coincides with the plane ACP. It is now easy to see that the planes ABM, BCN, and ACP have a common point X, and it satisfies

$$\frac{x_1}{S_1} = \frac{x_2}{S_2} = \frac{x_3}{S_3} = \frac{x_4}{S_4}.$$

Conversely, if the latter equalities hold, then X coincides with the intersection point of the planes ABM, BCN, and ACP.

(b) *Hint.* Use the Cauchy–Schwarz inequality as in the solution of Problem 1.2.5.

(c) Let L_1, L_2, L_3, and L_4 be the orthogonal projections of L onto the corresponding faces of the tetrahedron and let X be the centroid of $L_1L_2L_3L_4$. Then Leibniz's formula gives

$$LL_1^2 + LL_2^2 + LL_3^2 + LL_4^2$$
$$= 4LX^2 + XL_1^2 + XL_2^2 + XL_3^2 + XL_4^2$$
$$\geq x_1^2 + x_2^2 + x_3^2 + x_4^2 \geq LL_1^2 + LL_2^2 + LL_3^2 + LL_4^2,$$

which shows that $X = L$.

2.2.14 Let X_1, X_2, X_3, and X_4 be arbitrary points on the faces BCD, ACD, ABD, and ABC, respectively, of the given tetrahedron $ABCD$. Denote by t the sum of squares of the edges of tetrahedron $X_1X_2X_3X_4$, and by X the centroid of this tetrahedron. It follows from Leibniz's formula for $\triangle X_1X_2X_3$ that

$$XX_1^2 + XX_2^2 + XX_3^2 = 3X_4'X^2 + X_4'X_1^2 + X_4'X_2^2 + X_4'X_3^2,$$

where X_4' is the centroid of $\triangle X_1X_2X_3$. Since $XX_4' = \frac{1}{3}XX_4$ and

$$X_4'X_1^2 + X_4'X_2^2 + X_4'X_3^2 = \frac{1}{3}(X_1X_2^2 + X_2X_3^2 + X_3X_1^2),$$

the first equality gives

$$3(XX_1^2 + XX_2^2 + XX_3^2) = XX_4^2 + X_1X_2^2 + X_2X_3^2 + X_3X_1^2.$$

One gets similar equalities for each of the triangles $X_2X_3X_4$, $X_1X_3X_4$, and $X_1X_2X_4$. Summing these yields

$$t = 4(XX_1^2 + XX_2^2 + XX_3^2 + XX_4^2).$$

Since $XX_i \geq x_i$, $1 \leq i \leq 4$, the above together with Problem 2.2.13 implies

$$t \geq 4(x_1^2 + x_2^2 + x_3^2 + x_4^2) \geq 4(LL_1^2 + LL_2^2 + LL_3^2 + LL_4^2),$$

where L is Lemoine's point of $ABCD$. Equality holds only when $X = L$.

2.2.15 We may assume that the edge length of the regular tetrahedron $ABCD$ is 1. Then its vertices are four of the vertices of a cube of edge length $\frac{1}{2}\sqrt{2}$; the edges of the tetrahedron are the diagonals of six faces of the cube (Fig. 168).

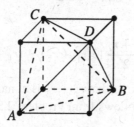

Figure 168.

The distance from a point X inside $ABCD$ to the diagonal of a face is not less than the distance from X to the face. Hence the desired minimum is equal to $\frac{3\sqrt{2}}{2}$. It is easy to see that it is attained only if X is the center of $ABCD$.

4.8 Malfatti's Problems

2.3.6 It is easy to observe that the Malfatti circles of an equilateral triangle of side length 1 have equal radii $r_1 = r_2 = r_3 = \frac{\sqrt{3}-1}{4}$.

The sum of their areas is $\frac{3\pi}{8}(2 - \sqrt{3})$. The incircle and the two small circles tangent to it and to two sides of the triangle (Fig. 169) have radii $\frac{1}{2\sqrt{3}}$, $\frac{1}{6\sqrt{3}}$, and $\frac{1}{6\sqrt{3}}$, respectively. So, the sum of their areas is equal to $\frac{11\pi}{108}$, and one checks that $\frac{11\pi}{108} > \frac{3\pi}{8}(2 - \sqrt{3})$.

2.3.7 As in Problem 2.3.1, one derives that it is enough to consider the case that the radii r_1 and r_2 of the two circles satisfy the conditions $r_1 + r_2 = 2 - \sqrt{2}$ and $0 \leq r_1, r_2 \leq \frac{1}{2}$.

Figure 169.

(a) The arithmetic mean–geometric mean inequality implies

$$r_1 r_2 \le \left(\frac{r_1 + r_2}{2}\right)^2 \le \left(1 - \frac{1}{\sqrt{2}}\right)^2,$$

i.e., $r_1 r_2$ is a maximum when $r_1 = r_2 = \frac{2-\sqrt{2}}{2}$.

(b) We have

$$r_1^3 + r_2^3 = \frac{r_1 + r_2}{2}\,[3(r_1^2 + r_2^2) - (r_1 + r_2)^2].$$

Since $r_1 + r_2 = 2 - \sqrt{2}$, it follows that $r_1^3 + r_2^3$ is a maximum when $r_1^2 + r_2^2$ is a maximum, and the solution follows from Problem 2.3.1.

2.3.8 *Hint.* Reduce the problem to the case that the two circles are tangent and are inscribed in two angles of the triangle (Fig. 170).

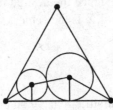

Figure 170.

Then use the arithmetic mean–geometric mean inequality.

2.3.9 Denote by a and b, $a \le b$, the side lengths of the given rectangle. If $b \ge 2a$, then one can put two circles of radius $\frac{a}{2}$ in the rectangle and the sum of their areas is a maximum (Fig. 171).

The interesting case is $a \le b < 2a$. As in Problem 2.3.1, one derives that it is enough to consider a pair of circles tangent to each other and inscribed in two opposite corners of the rectangle (Fig. 172).

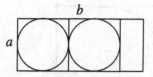

Figure 171.

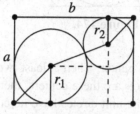

Figure 172.

Let r_1 and r_2 be their radii. Then it is easy to show that $r_1 + r_2 = a + b - \sqrt{2ab}$, and as in Problem 2.3.1 one finds that the sum of the areas of the two circles is a maximum when $r_1 = \frac{a}{2}$ and $r_2 = \frac{a}{2} + b - \sqrt{2ab}$ (Fig. 173).

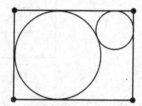

Figure 173.

2.3.10 It follows from Problem 2.3.3 that the required square has side length $3 + 2\sqrt{2}$.

2.3.11 *Hint.* Show that a square of side length $3 + 2\sqrt{2}$ contains three nonintersecting circles of radii 1, $\sqrt{2}$, and 2. Then the previous problem implies that such a square gives the solution to the problem.

2.3.12 *Answer.* $11\sqrt{3}$. *Hint.* Use Problem 2.3.4.

2.3.13 *Hint.* The solution is given by the incircle of the square and two circles inscribed in its angles and tangent to the incircle (Fig. 174).

To prove this proceed as in the solution of Problem 2.3.5 using Problem 2.3.3.

2.3.14 Assume that 5 nonintersecting unit circles are contained in a square of side a. Then $a \geq 2$ and the centers of the circles are contained in a square of side $a - 2$ (Fig. 175). Divide the latter square into 4 smaller squares of side $\frac{a-2}{2}$ using

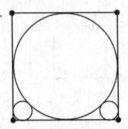

Figure 174.

two perpendicular lines through its center. Then at least two of the 5 centers lie in the same small square. If O_1 and O_2 are these centers, then $O_1O_2 \le \frac{a-2}{2}\sqrt{2}$. On the other hand, $O_1O_2 \ge 2$, since the unit circles have no common interior points. Hence $\frac{a-2}{2}\sqrt{2} \ge O_1O_2 \ge 2$, which gives $a \ge 2 + 2\sqrt{2}$.

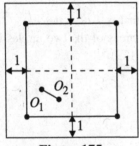

Figure 175.

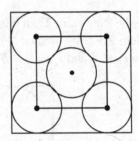

Figure 176.

On the other hand, it is easy to see (Fig. 176) that a square of side length $2+2\sqrt{2}$ contains 5 nonintersecting unit circles.

The answer is $2 + 2\sqrt{2}$.

2.3.15 *Hint.* It is enough to consider the case that the two balls are inscribed in two opposite trihedral angles of the cube and are tangent to each other.

2.3.16. *Hint.* Use the argument from the solution of Problem 2.3.3.

2.3.17 *Hint.* Use the argument from the solution of Problem 2.3.14.

4.9 Extremal Combinatorial Geometry Problems

2.4.6 Assume that $\triangle ABC$ is cut into n triangles satisfying the conditions of the problem (Fig. 177). Denote by v the number of all vertices in the net obtained in this way (including A, B, and C), and by k the number of segments issuing from one vertex. The sum of all angles in triangles from the net with vertices at a given point X is $360°$ (if $X = A$, B, or C, we also include the exterior angles in the sum; the sum of the three exterior angles is $3 \cdot 360° - 180° = 900°$). Thus, the sum of all angles in triangles of the net is $v \cdot 360° - 900°$. On the other hand, the same sum is equal to $n \cdot 180°$, so $n \cdot 180° = v \cdot 360° - 900°$, which implies $2v = n + 5$.

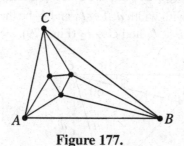

Figure 177.

The total number of segments in the net is $\frac{kv}{2}$, while the number of regions into which these segments divide the plane (counting the exterior of $\triangle ABC$ as well) is $n + 1$. Thus, $3(n + 1) = kv$, i.e., $n = \frac{kv}{3} - 1$. This and $2v = n + 5$ imply

$$n = \frac{k}{6}(n + 5) - 1 = \frac{nk}{6} + \frac{5k}{6} - 1,$$

i.e., $n = \frac{5k-6}{6-k}$. It is now easy to see that the only possible values for k are 2, 3, 4, and 5, and the corresponding values for n are 1, 3, 7, and 19. Thus $n \leq 19$. The case $n = 19$ is possible, as shown in Fig. 178.

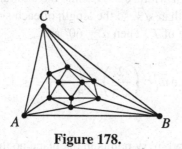

Figure 178.

2.4.7 Using induction on n, it is not difficult to show that n lines divide the plane into not more than $p(n) = \frac{n(n+1)}{2} + 1$ parts, where exactly $p(n)$ parts are obtained when any two lines intersect and no three lines intersect at one point. Next, using induction again, one shows that n planes divide the space into not more than $q(n) = \frac{n^3+5n+6}{6}$ parts, and one gets exactly $q(n)$ parts when any two of the planes intersect, no three of them have a common line, and no four of the planes have a common point. Since $q(12) = 299 < 300 < 378 = q(13)$, in order to cut the space into at

least 300 parts, one needs 13 planes. It is now easy to see that the same number of
planes are necessary to cut a cube into at least 300 parts.

2.4.8 Let ABC be an equilateral triangle of area 1. For the lengths a and h of its
side and altitude we have $a = \frac{2}{\sqrt[4]{3}}$ and $h = \sqrt[4]{3}$. Assume that triangle ABC is
contained in a horizontal strip with width d. Let ℓ_1 and ℓ_2 be the boundary lines of
the strip. We may assume that $B \in \ell_1$ and $C \in \ell_2$ (Fig. 179).

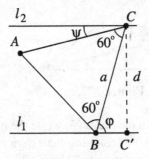

Figure 179.

Let φ and ψ be the angles between BC and ℓ_1 and AC and ℓ_2, respectively.
Then $\varphi = 60° + \psi \geq 60°$, so $d = CC' = a \sin \varphi \geq a \sin 60° = h$, where equality
holds precisely when $d = h$ and $A \in \ell_2$. In other words, the minimal width of a
strip containing triangle ABC is $\sqrt[4]{3}$.

Next, assume that T is an arbitrary triangle of area 1. We will show that T is
contained in a horizontal strip of width $\sqrt[4]{3}$. Assume the contrary. Then the length
of each altitude of T is greater than $\sqrt[4]{3}$, so the length of each side of T is less than
$\frac{2}{\sqrt[4]{3}}$. Let α be the smallest angle of T. Then $\alpha \leq 60°$ and

$$[T] = \frac{bc}{2} \sin \alpha < \left(\frac{2}{\sqrt[4]{3}}\right)^2 \cdot \frac{1}{4} = \frac{1}{\sqrt{3}} < 1,$$

a contradiction.

2.4.9 Let $A_1, A_2, \ldots, A_n$ be n arbitrary points in the plane no three of which lie on
a line. We will show that $\alpha \leq \frac{180°}{n}$. There exist two points, say $A_1 \neq A_2$, such
that all the other points lie in one of the half-planes determined by the line $A_1 A_2$.
Choose a point A_3 of the given ones such that $\angle A_1 A_2 A_3$ is a maximum; then all
the other points are contained in this angle. Moreover, $\angle A_1 A_2 A_3 \geq \alpha(n-2)$, since
the angle between any two successive rays $A_2 A_i$ is not less than α (Fig. 180).

Then we choose a point A_4 such that $\angle A_2 A_3 A_4$ is a maximum, etc. Clearly
we have $\angle A_2 A_3 A_4 \geq \alpha(n-2)$, $\angle A_3 A_4 A_5 \geq \alpha(n-2)$, etc. Since the number of the given points is n, there exists a minimal number $m \leq n$ such that

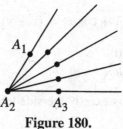

Figure 180.

$A_{m+1} \in \{A_1, A_2, \ldots, A_{m-1}\}$ (clearly $A_{m+1} \neq A_m$), that is, $\angle A_{m-1}A_mA$ is a maximum for $A = A_i$ for some $1 \leq i \leq m - 1$. If $i \neq 1$, then A_1 lies in the angle $A_{m-1}A_mA_i$, a contradiction.

Thus, $i = 1$ and any of the angles of the convex polygon $A_1A_2\ldots A_m$ is not less than $\alpha(n - 2)$. Hence $180°(m - 2) \geq m\alpha(n - 2)$. This implies

$$\alpha \leq \frac{180°(m - 2)}{m(n - 2)} = \frac{180°}{n - 2}\left(1 - \frac{2}{m}\right) \leq \frac{180°}{n - 2}\left(1 - \frac{2}{n}\right) = \frac{180°}{n}.$$

It is easy to see that if $A_1, A_2, \ldots, A_n$ are the vertices of a regular n-gon, then $\alpha = \frac{180°}{n}$ (Fig. 181). Hence the largest possible value of α is $\frac{180°}{n}$.

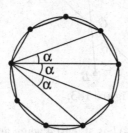

Figure 181.

2.4.10 *Hint.* Let $ABCD$ be the given rectangle, where $AB = 4$ and $BC = 3$. First show that it is enough to consider the case $A_1 = A$, $A_2 \in AB$, $A_3 = C$, $A_4 \in CD$ (Fig. 182). Then show that the desired maximum is achieved precisely when $A_1A_2A_3A_4$ is a rhombus. In this case $A_1A_2 = \frac{25}{8}$.

2.4.11 Let O be the intersection point of the segments. There exists a side AB of the $2n$-gon such that $\angle AOB \geq \frac{180°}{n}$ and $AO + OB \geq 1$ (Fig. 183). Set $x = AO$ and $y = OB$. Then $x + y \geq 1$ and the law of cosines for $\triangle AOB$ gives

$$AB^2 = x^2 + y^2 - 2xy\cos\alpha = (x + y)^2 - 2xy(1 + \cos\alpha)$$

$$\geq (x+y)^2 - \frac{(x+y)^2}{2}(1+\cos\alpha) = (x+y)^2 \cdot \frac{1-\cos\alpha}{2}$$

$$\geq \frac{1-\cos\frac{180°}{n}}{2} = \sin^2\frac{90°}{n}.$$

Thus, $AB \geq \sin\frac{90°}{n}$, and the latter is exactly the side length of a regular $2n$-gon inscribed in a circle with diameter 1.

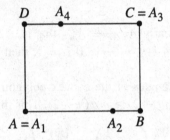

Figure 182.

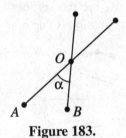

Figure 183.

2.4.12 If a convex n-gon (with $n \geq 4$) has at least 4 acute angles, then their exterior angles will be obtuse, so their sum will be greater than $360°$. However, the sum of all exterior angles $\alpha_1, \alpha_2, \ldots, \alpha_n$ (Fig. 184) of the n-gon is $n \cdot 180° - (n-2)180° = 360°$, a contradiction.

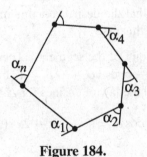

Figure 184.

2.4.13

(a) Let $p_1, p_2, \ldots, p_n$ be rays in space issuing from a point O, and assume that the angle between any two rays p_i and p_j is obtuse. For each i let A_i be the point on p_i with $OA_i = 1$. Then $\overrightarrow{OA_i} \cdot \overrightarrow{OA_j} < 0$ for all $i \neq j$. Consider a coordinate system $Oxyz$ in space such that $A_1 = (1, 0, 0)$, $A_2 = (x_2, y_2, 0)$, and $A_k = (x_k, y_k, z_k)$, $3 \leq k \leq n$, where $y_2 > 0$ and $z_3 > 0$. Then $x_i = \overrightarrow{OA_i} \cdot \overrightarrow{OA_1} < 0$ for $i > 1$. Moreover, $x_i x_2 + y_i y_2 = \overrightarrow{OA_i} \cdot \overrightarrow{OA_2} < 0$ for $i > 2$, which, combined with $x_i < 0$, $x_2 < 0$, and $y_2 > 0$, gives $y_i < 0$ for all $i > 2$. Finally, $x_i x_3 + y_i y_3 + z_i z_3 = \overrightarrow{OA_i} \cdot \overrightarrow{OA_3} < 0$ implies $z_i < 0$ for $i > 3$.

Now if $n > 4$, then for A_4 and A_5 we have $x_4 x_5 > 0$, $y_4 y_5 > 0$, and $z_4 z_5 > 0$, which is a contradiction to $\overrightarrow{OA_4} \cdot \overrightarrow{OA_5} < 0$. Thus we must have $n \leq 4$.

That $n = 4$ is possible is seen by considering the rays issuing from the center O of a regular tetrahedron and passing through its vertices (Fig. 185).

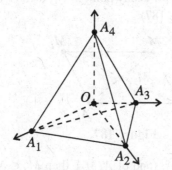

Figure 185.

(b) *Answer.* 6.

2.4.14

(a) *Answer.* 4 points.

(b) Let $A_1, A_2, \ldots, A_n$ be points in space such that any of the angles $A_i A_j A_k$ does not exceed $90°$. We will show that $n \leq 8$. Given two points A_i and A_j, denote by Π_{ij} the strip between the planes passing through A_i and A_j and perpendicular to the line $A_i A_j$ (Fig. 186). Clearly Π_{ij} coincides with the set of points M in space such that $\angle M A_i A_j \leq 90°$ and $\angle M A_j A_i \leq 90°$. Let N be the convex hull of the set $\{A_1, A_2, \ldots, A_n\}$. Then N is a convex polyhedron and each A_i lies inside or on the boundary of N. Moreover, N is contained in the intersection of all strips Π_{ij}.

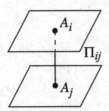

Figure 186.

Given $i = 1, \ldots, n$, denote by N_i the polyhedron obtained from N by the translation along the vector $\overrightarrow{A_1 A_i}$. Let N' be the image of N under the dilation φ with center A_1 and ratio 2. Since $N \subset \Pi_{ij}$, the polyhedron obtained by a translation along $\overrightarrow{A_i A_j}$ has no common interior points with N. Since N_j is the image of N_i under the translation of N_i along $\overrightarrow{A_i A_j}$, the previous remark shows that N_i and N_j have no common interior points when $i \neq j$. Also notice that $N_i \subset N'$ for all i. Indeed, if M_i is any point in N_i, then $\overrightarrow{A_1 M_i} = \overrightarrow{A_1 M} + \overrightarrow{A_1 A_i}$ for some point M of N (Fig. 187).

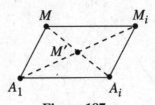

Figure 187.

If M' is the midpoint of $M A_i$ (and of $A_1 M_i$), then $M' \in N$ and $\varphi(M') = M_i$. So $M_i \in N' = \varphi(N)$, and therefore $N_i \subset N'$ for any i. Consequently,

$$n \cdot \text{Vol}(N) = \text{Vol}(N_1) + \text{Vol}(N_2) + \cdots + \text{Vol}(N_n) \leq \text{Vol}(N') = 8\text{Vol}(N),$$

so $n \leq 8$. Clearly the vertices $A_1, A_2, \ldots, A_8$ of any cube satisfy the requirements of the problem.

2.4.15 Let O be the center of the disk and let $A_1, A_2, \ldots, A_n$ be points in the disk such that $A_i A_j > 1$ for $i \neq j$ (Fig. 188).

We may assume that these points are ordered clockwise since no two of them lie on the same radius. Set $\alpha_i = \angle A_i O A_{i+1}, 1 \leq i \leq n$ $(A_{n+1} = A_1)$. Then $\alpha_i > 60°$ since $A_i A_{i+1}$ is the largest side of $\triangle A_i O A_{i+1}$. Hence

$$360° = \alpha_1 + \alpha_2 + \cdots + \alpha_n > n \cdot 60°$$

and therefore $n \leq 5$. To prove that the desired number is 5, take five points A_1, A_2, A_3, A_4, A_5 that are sufficiently close to the vertices of a regular pentagon inscribed in a unit circle (Fig. 189).

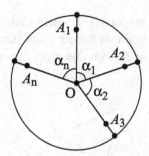

Figure 188.

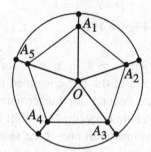

Figure 189.

2.4.16 Let $A_1 A_2 \ldots A_n$ be an arbitrary convex n-gon. The diagonals through A_1 cut it into $n - 2$ triangles. That is why the desired number of points is not less than $n - 2$. A distribution of $n - 2$ points in a convex n-gon satisfying the requirements of the problem is shown in Fig. 190.

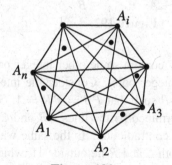

Figure 190.

2.4.17 Let Π be a rectangle with side lengths a and b, $a \le b$, that has the required property (henceforth we denote this property by (∗)). Then clearly $a \ge 1$. In what follows we consider only rectangles Π with $a \ge 1$.

Assume that Π does not have property (∗). Then there is a position of Π in the plane for which Π does not contain an integer point (i.e., a point with integer coordinates). Consider a position of Π with this property, and extend its sides of length a (Fig. 191).

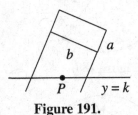

Figure 191.

There exists a line with equation $y = k$ or $x = k$ for some integer k that intersects the strip obtained. The part of the line contained in the strip has length at least b. Since $b \geq a \geq 1$, this part contains an integer point P. We may assume that P is the closest integer point to Π in the strip. Now shift Π in the strip keeping its sides parallel to their initial positions until one of the sides of Π with length b passes through P (Fig. 192).

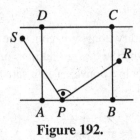

Figure 192.

Notice that the side AB may contain some other integer points. However, the rest of Π does not contain an integer point. Consider the integer points S and R on the coordinate lines through P such that $PS = PR = 1$. Since $AD = a \geq 1$, R and S lie inside the strip determined by the lines AB and CD. If R lies on the line AB, then S will be in Π, a contradiction. In the same way one observes that S does not lie on AB. Hence both S and R lie outside Π, which gives $b = AB < RS = \sqrt{2}$.

Conversely, let $1 \leq a \leq b < \sqrt{2}$. Then it is easy to see that there is a position of Π for which Π does not contain an integer point (Fig. 193).

Thus Π has property (∗) if and only if $a \geq 1, b \geq a$, and $b \geq \sqrt{2}$. It is clear now that the minimum area of Π is equal to $\sqrt{2}$.

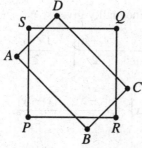

Figure 193.

4.10   Triangle Inequality

3.1.1 Assume that the triangle ABC has sides of lengths 1. Set $AX = x$ and $BY = y$, where $0 \leq x, y \leq 1$. Let X_1 and X_2 be the orthogonal projections of X on AB and BC, and let Y_1 and Y_2 be the orthogonal projections of Y on AB and AC (Fig. 194). Then $AX_1 = \frac{x}{2}$ since $\angle AXX_1 = 30°$. Analogously, $BY_1 = \frac{y}{2}$, $CX_2 = \frac{1-x}{2}$, $CY_2 = \frac{1-y}{2}$. Hence

$$S(X, Y) = X_1Y_1 + XY_2 + YX_2$$
$$= 1 - \frac{x+y}{2} + \left| 1 - x - \frac{1-y}{2} \right| + \left| 1 - y - \frac{1-x}{2} \right|$$
$$= 1 - \frac{x+y}{2} + \left| \frac{1}{2} + \frac{x}{2} - y \right| + \left| \frac{1}{2} + \frac{y}{2} - x \right|.$$

It is clear that the minimum of $S(X, Y)$ is equal to 0, and it is attained if and only if $X = Y = C$.

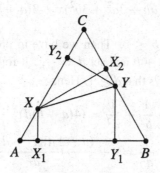

Figure 194.

On the other hand, the triangle inequality gives

$$\left|\frac{1}{2} + \frac{x}{2} - y\right| = \left|\frac{1-y}{2} + \frac{x-y}{2}\right| \le \frac{1-y}{2} + \frac{|x-y|}{2}.$$

Analogously

$$\left|\frac{1}{2} + \frac{y}{2} - x\right| \le \frac{1-x}{2} + \frac{|x-y|}{2}.$$

Hence

$$S(X, Y) \le 1 - \frac{x+y}{2} + \frac{1-x}{2} + \frac{1-y}{2} + |x-y|$$

$$= 2 + |x - y| - (x + y) \le 2.$$

Thus the maximum of $S(X, Y)$ is equal to 2 and it is attained if and only if $X = A, Y = C$ or $X = C, Y = B$.

3.1.2 We first prove that $k \ge \frac{1+\sqrt{5}}{2}$. Indeed, let m be an arbitrary real number such that $1 \le m < \frac{1+\sqrt{5}}{2}$. Then $1 + m > m^2$, which shows that there exists a triangle with side lengths $1, m, m^2$. Hence $k > \min\left(\frac{m}{1}, \frac{m^2}{m}, \frac{m^2}{1}\right) = m$, implying $k \ge \frac{1+\sqrt{5}}{2}$. Conversely, let $k \ge \frac{1+\sqrt{5}}{2}$ and suppose that the assertion is not true. Then there exists a triangle with side lengths $a \ge b \ge c$ such that $\frac{a}{b} \ge k$ and $\frac{b}{c} \ge k$. We derive that $b \le \frac{a}{k}$ and $c \le \frac{b}{k} \le \frac{a}{k^2}$. Hence $b + c \le a\left(\frac{1}{k} + \frac{1}{k^2}\right) \le a$, a contradiction.

Thus the least possible value of k is equal to $\frac{1+\sqrt{5}}{2}$.

3.1.3 We have to find the greatest real number k such that for any $a, b, c > 0$ with $a + b \le c$, we have $kabc \le a^3 + b^3 + c^3$. First take $b = a$ and $c = 2a$. Then $2ka^3 \le 10a^3$, i.e., $k \le 5$. Conversely, let $k = 5$. Set $c = a + b + x$, where $x \ge 0$. Then

$$a^3 + b^3 + c^3 - 5abc$$

$$= 2(a+b)(a-b)^2 + (ab + 3a^2 + 3b^2)x + 3(a+b)x^2 + x^3 \ge 0.$$

3.1.4 We may assume that $a \le b \le c$. Then we have to prove that $c < a + b$. Suppose the contrary, i.e., $c \ge a + b$ and set $d = \frac{1}{4ab}$. It follows that $d^2 \ge c^2 \ge (a+b)^2 \ge 4ab = \frac{1}{d}$, which shows that $d \ge 1$. Hence

$$\frac{1}{a^2} + \frac{1}{b^2} + \frac{1}{c^2} \ge \frac{1}{a^2} + \frac{1}{b^2} + \frac{1}{d^2} = (4(a+b)d)^2 - 8d + \frac{1}{d^2}$$

$$\ge 8d + \frac{1}{d^2} = 9 + \frac{(d-1)(8d^2 - d - 1)}{d^2} \ge 9,$$

a contradiction.

3.1.5

(a) The given inequality follows from the identity

$$a^3 + b^3 + c^3 - (a + b + c)(ab + bc + ca)$$
$$= a^2(a - b - c) + b^2(b - c - a) + c^2(c - a - b) - 3abc$$

and the triangle inequality.

(b) If $a = b = 1$, then

$$k > \frac{2 + c^3}{(2 + c)(1 + 2c)} = 1 - \frac{5t^2 + 2t - 1}{2t^3 + 5t^2 + 2t} = 1 - f(t),$$

where $t = \frac{1}{c}$. Since $\lim_{t \to \infty} f(t) = 0$, it follows that $k \geq 1$. Using (a) we deduce that the least value of k is equal to 1.

3.1.6 Let ABC be a triangle with $AB = c, BC = a, CA = b$. Since $p + q + r = 0, pqr \neq 0$, it follows that two of these numbers, say p and r, have the same sign. Then $pr > 0$, and the law of cosines implies that

$$a^2 pq + b^2 qr + c^2 rp = a^2 pq + b^2 qr + rp(a^2 + b^2 - 2ab \cos C)$$
$$= -a^2 p^2 - b^2 r^2 - 2abrp \cos C$$
$$= -(ap - br)^2 - 2abrp(1 + \cos C) < 0.$$

Conversely, setting $p = b, q = c, r = -(b + c)$ we get $bc(a^2 - (b + c)^2) < 0$, i.e., $a < b + c$. Analogously $b < c + a, c < a + b$, and therefore a, b, c are the side lengths of a triangle.

3.1.7 To show (i) implies (ii), note that

$$a^2 x + b^2 y + c^2 z \geq (a^2 x + b^2 y + c^2 z) \left(\frac{1}{x} + \frac{1}{y} + \frac{1}{z} \right)$$
$$\geq (a + b + c)^2 > d^2,$$

where we have used the Cauchy–Schwarz inequality and the triangle inequality.

To show (ii) implies (i), first note that if $x \leq 0$, we may take a quadrilateral with side lengths $a = n, b = 1, c = 1, d = n$ and get $y + z > n^2(1 - x)$, a contradiction for large n. Thus, $x > 0$ and similarly $y > 0, z > 0$. Now use a quadrilateral with side lengths $\frac{1}{x}, \frac{1}{y}, \frac{1}{z}$ and $\frac{1}{x} + \frac{1}{y} + \frac{1}{z} - \frac{1}{n}$, where n is large. We then have

$$\frac{x}{x^2} + \frac{y}{y^2} + \frac{z}{z^2} > \left(\frac{1}{x} + \frac{1}{y} + \frac{1}{z} - \frac{1}{n} \right)^2,$$

and taking the limit as $n \to \infty$ we get

$$\frac{1}{x} + \frac{1}{y} + \frac{1}{z} \geq \left(\frac{1}{x} + \frac{1}{y} + \frac{1}{z}\right)^2.$$

Hence $\frac{1}{x} + \frac{1}{y} + \frac{1}{z} \leq 1$.

4.11 Selected Geometric Inequalities

3.2.1

(i) Set $x = a + b - c > 0$, $y = b + c - a > 0$, $z = c + a - b > 0$. Then $a = \frac{y+z}{2}$, $b = \frac{x+z}{2}$, $c = \frac{x+y}{2}$ and we have to prove that $(x+y)(y+z)(z+x) \geq 8xyz$. This follows by multiplying the inequalities $x + y \geq 2\sqrt{xy}$, $y + z \geq 2\sqrt{yz}$, $z + x \geq 2\sqrt{zx}$.

(ii) Let F be the area of the triangle. Then $abc = 4RF$, $F = sr$, and $F^2 = s(s-a)(s-b)(s-c)$. Hence the inequality (i) can be written as $\frac{8F^2}{s} \leq 4RF$, which is equivalent to $R \geq 2r$.

There is a nice geometric proof of Euler's inequality based on the "obvious" observation that the incircle is the smallest circle having common points with the three sides of a triangle. Indeed, let A_1, B_1, and C_1 be the midpoints of the sides of a triangle ABC. Then the circumradius of triangle $A_1B_1C_1$ is equal to $\frac{R}{2}$ and therefore $\frac{R}{2} \geq r$.

(iii) We have

$$r^2 = \frac{(s-a)(s-b)(s-c)}{s}$$

$$= \frac{s^3 - s^2(a+b+c) + s(ab+bc+ca) - abc}{s}$$

$$= -s^2 + ab + bc + ca - 4Rr.$$

Hence $\sigma_2 = ab + bc + ca = s^2 + r^2 + 4Rr$. Set $\sigma_1 = a + b + c = 2s$ and $\sigma_3 = abc = 4srR$. Then a direct computation shows that

$$(a-b)^2(b-c)^2(c-a)^2 = \sigma_1^2\sigma_2^2 - 4\sigma_2^3 - 4\sigma_1^3\sigma_3 + 18\sigma_1\sigma_2\sigma_3 - 27\sigma_3^2$$

$$= -4r^2[(s^2 - 2R^2 - 10Rr + r^2)^2 - 4R(R-2r)^3].$$

Thus $(s^2 - 2R^2 - 10Rr + r^2)^2 - 4R(R-2r)^3 \leq 0$, which is equivalent to $|s^2 - 2R^2 - 10Rr + r^2| \leq 2(R-2r)\sqrt{R(R-2r)}$.

Remark. The inequalities (ii) and (iii) are also sufficient for the existence of a triangle with semiperimeter s, circumradius R, and inradius r. Moreover, Blundon [5] has proved that (iii) is the strongest possible inequality of the form $f(R, r) \leq s^2 \leq F(R, r)$, where $f(R, r)$ and $F(R, r)$ are homogeneous real functions, with simultaneous equality only for equilateral triangles. For the history of the fundamental inequality we refer the reader to [15].

(iv) The inequality $R \geq 2r$ together with (iii) implies that

(1) $$16Rr - 5r^2 \leq s^2 \leq 4R^2 + 4Rr + 3r^2.$$

Now using the indentity

$$a^2 + b^2 + c^2 = 4s^2 - 2(ab + bc + ca) = 2s^2 - 2r^2 - 8Rr$$

one gets

$$24Rr - 12r^2 \leq a^2 + b^2 + c^2 \leq 8R^2 + 4r^2.$$

(v) The given inequalities follow from (1) since $R \geq 2r$ implies that $16Rr - 5r^2 \geq 27r^2$ and $(2R + (3\sqrt{3} - 4)r)^2 \geq 4R^2 + 4Rr + 3r^2$.

3.2.2 Set $\frac{AM}{MC} = x$, $\frac{CN}{NB} = y$, and $\frac{ML}{LN} = z$ (Fig. 195).

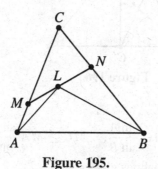

Figure 195.

Then $[MLC] = \frac{1}{x}S_1$, $[NLC] = yS_2$ and therefore $S_1 = xyzS_2$ since $[MLC] = z[NLC]$. Hence

$$[MNC] = [MLC] + [NLC] = z(y + 1)S_2$$

and we get

$$S = \frac{AC}{MC} \cdot \frac{BC}{NC}[MNC] = (1 + x)(1 + y)(1 + z)S_2.$$

Thus we have to prove the inequality

$$(1 + x)(1 + y)(1 + z) \geq (1 + \sqrt[3]{xyz})^3.$$

It follows by the arithmetic mean–geometric mean inequality since

$$(1 + x)(1 + y)(1 + z) = 1 + x + y + z + xy + yz + zx + xyz$$

$$\geq 1 + 3\sqrt[3]{xyz} + 3\sqrt[3]{(xyz)^2} + xyz$$

$$= (1 + \sqrt[3]{xyz})^3.$$

3.2.3

(i) Since M, B', A, C' are cyclic points (Fig. 196) we have $B'C' = MA \sin A$.
The length of the orthogonal projection of the segment $B'C'$ on the line BC is
equal to $MB' \cos(90° - C) + MC' \cos(90° - B) = MB' \sin C + MC' \sin B$.
Hence

$$MA \geq MB' \frac{\sin C}{\sin A} + MC' \frac{\sin B}{\sin A}.$$

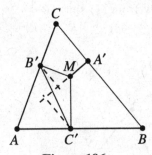

Figure 196.

Analogously,

$$MB \geq MA' \frac{\sin C}{\sin B} + MC' \frac{\sin A}{\sin B}, \quad MC \geq MA' \frac{\sin B}{\sin C} + MB' \frac{\sin A}{\sin C}.$$

Summing up the above inequalities gives

$$MA + MB + MC \geq MA' \left(\frac{\sin B}{\sin C} + \frac{\sin C}{\sin B} \right)$$

$$+ MB' \left(\frac{\sin A}{\sin C} + \frac{\sin C}{\sin A} \right) + MC' \left(\frac{\sin A}{\sin B} + \frac{\sin B}{\sin A} \right).$$

This implies the desired inequality since

$$\frac{\sin B}{\sin C} + \frac{\sin C}{\sin B} \geq 2, \quad \frac{\sin A}{\sin C} + \frac{\sin C}{\sin A} \geq 2, \quad \frac{\sin A}{\sin B} + \frac{\sin B}{\sin A} \geq 2.$$

(ii) Denote by A'', B'', C'' and A''', B''', C''' the points on the rays MA', MB', MC' and MA, MB, MC such that $MA'' = \frac{1}{MA'}$, $MB'' = \frac{1}{MB'}$, $MC'' = \frac{1}{MC'}$, and $MA''' = \frac{1}{MA}$, $MB''' = \frac{1}{MB}$, $MC''' = \frac{1}{MC}$ (Fig. 197).

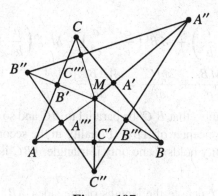

Figure 197.

Then triangles $MB'A$ and $MB''A'''$ are similar since

$$\frac{MB'}{MA} = \frac{MA'''}{MB''}.$$

Hence $\angle MA'''B'' = \angle MB'A = 90°$. Analogously $\angle MA'''C'' = \angle MC'A = 90°$ and therefore the points B'', A''', and C'' are collinear. Thus A''', B''', C''' are the orthogonal projections of M on the lines $B''C''$, $A''C''$, $A''B''$, respectively. Now applying (i) to triangle $A''B''C''$ and the point M, we get

$$\frac{1}{MA'} + \frac{1}{MB'} + \frac{1}{MC'} = MA'' + MB'' + MC''$$

$$\geq 2(MA''' + MB''' + MC''')$$

$$= 2\left(\frac{1}{MA} + \frac{1}{MB} + \frac{1}{MC}\right).$$

3.2.4 Let $BC = a$, $AC = b$, $AB = c$. Using the same notation as in the solution of Problem 3.2.3, we have

(1) $$MA \sin A \geq MB' \sin C + MC' \sin B.$$

Multiplying by $2R$ and using the law of sines, (1) becomes

$$aMA \geq cMB' + bMC'.$$

Likewise, we have $bMB \geq aMC' + cMA'$ and $cMC \geq bMA' + aMB'$. Using these inequalities, we obtain

$$\frac{MA}{a^2} + \frac{MB}{b^2} + \frac{MC}{c^2}$$

$$\geq MA'\left(\frac{b}{c^3} + \frac{c}{b^3}\right) + MB'\left(\frac{c}{a^3} + \frac{a}{c^3}\right) + MC'\left(\frac{a}{b^3} + \frac{b}{a^3}\right)$$

$$\geq \frac{2MA'}{bc} + \frac{2MB'}{ca} + \frac{2MC'}{ab} = \frac{4[ABC]}{abc} = \frac{1}{R}.$$

Equality in the first step requires that $B'C'$ be parallel to BC and so on. This occurs if and only if M is the circumcenter of ABC. Equality in the second step requires that $a = b = c$. Thus, equality holds if and only if triangle ABC is equilateral and M is its center.

3.2.5 Let a, b, c, d, e, and f denote the lengths of the sides AB, BC, CD, DE, EF, and FA, respectively. Note that the opposite angles of the hexagon are equal $(\angle A = \angle D, \angle B = \angle E, \angle C = \angle F)$.

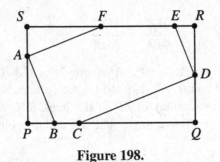

Figure 198.

Draw perpendiculars as follows: $AP \perp BC$, $AS \perp EF$, $DQ \perp BC$, $DR \perp EF$ (Fig. 198). Then $PQRS$ is a rectangle and $BF \geq PS = QR$. Therefore $2BF \geq PS + QR$, and so

$$2BF \geq (a \sin B + f \sin C) + (c \sin C + d \sin B).$$

Similarly,

$$2DB \geq (c \sin A + d \sin B) + (e \sin B + f \sin A),$$

$$2FD \geq (e \sin C + d \sin A) + (a \sin A + b \sin C).$$

Next, the circumradii of triangles FAB, BCD, and DEF are related to BF, DB, and FD as follows:

$$R_A = \frac{BF}{2 \sin A}, \quad R_C = \frac{DB}{2 \sin C}, \quad R_E = \frac{FD}{2 \sin B}.$$

We obtain, therefore,

$$4(R_A + R_C + R_E) \geq a\left(\frac{\sin B}{\sin A} + \frac{\sin A}{\sin B}\right) + b\left(\frac{\sin B}{\sin C} + \frac{\sin C}{\sin B}\right) + \cdots$$
$$\geq 2(a + b + \cdots) = 2P,$$

and so $R_A + R_C + R_E \geq P/2$, as required. Equality holds iff $\angle A = \angle B = \angle C$ and $BF \perp BC, \ldots$, that is, iff the hexagon is regular.

3.2.6 *First solution.* We may assume that all four points are different; otherwise, the given inequality is obvious. Let B', C', D' be the points on the rays AB, AC, and AD respectively, such that $AB' \cdot AB = AC' \cdot AC = AD' \cdot AD = 1$. Then $\frac{AB}{AC} = \frac{AC'}{AB'}$, which shows that triangles ABC and $AB'C'$ are similar. Hence

$$B'C' = \frac{BC}{AB \cdot AC}.$$

Analogously,

$$C'D' = \frac{CD}{AC \cdot AD} \quad \text{and} \quad B'D' = \frac{BD}{AB \cdot AD}.$$

Now the triangle inequality $B'C' + C'D' \geq B'D'$ implies that $AB \cdot CD + AD \cdot BC \geq AC \cdot BD$.

Equality is obtained if and only if the quadrilateral $ABCD$ is cyclic.

Second solution. Let a, b, c, d be the complex numbers representing the points A, B, C, D, respectively. Then the triangle inequality implies that

$$AC \cdot BD = |a - c| \cdot |b - d| = |(a - c)(b - d)|$$
$$= |(a - b)(c - d) + (a - d)(b - c)|$$
$$\leq |a - b| \cdot |c - d| + |a - d| \cdot |b - c| = AB \cdot CD + AD \cdot BC.$$

3.2.7 Let us set $AC = a, CE = b, AE = c$. Applying Ptolemy's inequality for the quadrilateral $ACEF$, we get

$$AC \cdot EF + CE \cdot AF \geq AE \cdot CF.$$

Since $EF = AF$, we have $\frac{FA}{FC} \geq \frac{c}{a+b}$. Similarly, $\frac{DE}{DA} \geq \frac{b}{c+a}$ and $\frac{BC}{BE} \geq \frac{a}{b+c}$. It follows that

(1) $$\frac{BC}{BE} + \frac{DE}{DA} + \frac{FA}{FC} \geq \frac{a}{b+c} + \frac{b}{c+a} + \frac{c}{a+b} \geq \frac{3}{2},$$

where the last inequality is left as an exercise to the reader.

For equality to occur we need (1) to be an equality and also we need an equality each time Ptolemy's inequality was used. The latter happens when the quadrilaterals $ACEF$, $ABCE$, $ACDE$ are cyclic, that is, when $ABCDEF$ is a cyclic hexagon. Also, for the equality in (1) we need $a = b = c$.

Hence equality occurs if and only if the hexagon is regular.

3.2.8 If O lies on AC, then $ABCD$, $AKON$, and $OLCM$ are similar, and $AC = AO + OC$ (Fig. 199). Hence $\sqrt{S} = \sqrt{S_1} + \sqrt{S_2}$.

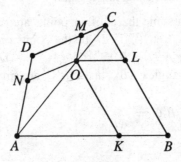

Figure 199.

If O does not lie on AC, we may assume that O and D are on the same side of AC. Denote the points of intersection of a line through O with BA, AD, CD, and BC by W, X, Y, and Z, respectively (Fig. 200).

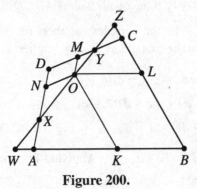

Figure 200.

Initially, let $W = X = A$. Then $\frac{OW}{OX} = 1$, while $\frac{OZ}{OY} > 1$. Rotate the line about O without passing through B, until $Y = Z = C$. Then $\frac{OW}{OX} > 1$, while $\frac{OZ}{OY} = 1$. Hence in some position during the rotation, we have $\frac{OW}{OX} = \frac{OZ}{OY}$. Fix the line there. Let T_1, T_2, P_1, P_2, Q_1, and Q_2 denote the areas of $KBLO$, $NOMD$, WKO, OLZ, ONX, and YMO, respectively. The desired result is equivalent to $T_1 + T_2 \geq 2\sqrt{S_1 S_2}$. Since triangles WBZ, WKO, and OLZ are similar, we have

$$\sqrt{P_1} + \sqrt{P_2} = \sqrt{P_1 + T_1 + P_2}\left(\frac{WO}{WZ} + \frac{OZ}{WZ}\right) = \sqrt{P_1 + T_1 + P_2},$$

which is equivalent to $T_1 = 2\sqrt{P_1 P_2}$. Similarly, $T_2 = 2\sqrt{Q_1 Q_2}$. Since $\frac{OW}{OZ} = \frac{OX}{OY}$, we have

$$\frac{P_1}{P_2} = \frac{OW^2}{OZ^2} = \frac{OX^2}{OY^2} = \frac{Q_1}{Q_2}.$$

Denote the common value of $\frac{Q_1}{P_1} = \frac{Q_2}{P_2}$ by k. Then

$$T_1 + T_2 = 2\sqrt{P_1 P_2} + 2\sqrt{Q_1 Q_2} = 2\sqrt{P_1 P_2}(1 + k)$$

$$= 2\sqrt{(1 + k)P_1(1 + k)P_2} = 2\sqrt{(P_1 + Q_1)(P_2 + Q_2)} \geq 2\sqrt{S_1 S_2}.$$

3.2.9 We first show that the result holds when F is a "digon," i.e., a polygon with only 2 sides. Let O be a point and AB a line segment. Set $OA = a, OB = b, AB = c$ and let the distance of O from the line AB be h. Treating the figure ABA as a two-sided polygon, we find that $D = a + b, P = 2c$ (this being the perimeter of the digon), and $H = 2h$. The inequality $D^2 \geq H^2 + P^2/4$ now takes the form $(a + b)^2 \geq 4h^2 + c^2$.

To prove this, we draw a line l through O parallel to AB, and let B_1 be the image of B under reflection in l (Fig. 201). Then $OA + OB = OA + OB_1 \geq AB_1$, i.e., $a + b \geq \sqrt{4h^2 + c^2}$, which is precisely the stated inequality. Note that equality holds iff $\angle OAB = \angle OBA$, i.e., iff $a = b$.

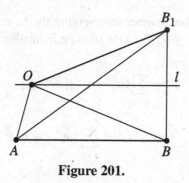

Figure 201.

Now let the polygon F be $P_1 P_2 \ldots P_n$, and let

$$d_i = OP_i, \quad p_i = P_i P_{i+1}, \quad h_i = \text{distance from } O \text{ to } P_i P_{i+1}.$$

(Here P_{n+1} is the same as P_1.) For each i, using the result proved above,

$$d_i + d_{i+1} \geq \sqrt{4h_i^2 + p_i^2}.$$

Summing these inequalities over $i = 1, 2, \ldots, n$, we obtain

$$2D \geq \sum_i \sqrt{4h_i^2 + p_i^2},$$

or after squaring,

$$4D^2 \geq \left(\sum_i \sqrt{4h_i^2 + p_i^2} \right)^2.$$

Now $4H^2 + P^2 = 4(\sum h_i)^2 + (\sum p_i)^2$, so it suffices to prove that

$$\sum_i \sqrt{4h_i^2 + p_i^2} \geq \sqrt{4 \left(\sum_i h_i \right)^2 + \left(\sum_i p_i \right)^2}.$$

Let v_i denote the vector with coordinates $(2h_i, p_i)$. Then the quantity on the left side is $\sum |v_i|$ and the quantity on the right side is $|\sum v_i|$, and the inequality follows by the triangle inequality.

Equality holds if and only if (a) the d_i's are all equal, say $d_i = r$ for all i, which means that the P_i's lie on the circle $C(O, r)$, and (b) the ratio h_i / p_i is the same for all i. Since each side of F is a chord of C, we have $h_i^2 + p_i^2/4 = r^2$ for all i, so the constancy of h_i / p_i implies that the h_i's are all equal, and likewise the p_i's. Thus equality holds if and only if F is a regular polygon and O is its circumcenter.

3.2.10 Denote by z_k the complex number representing the point A_k, $1 \leq k \leq 2n$, and set $w_k = z_{n+k} - z_k$, $1 \leq k \leq n$. Then the triangle inequality gives

$$\sum_{k=1}^{n} (A_k A_{k+1} + A_{n+k} A_{n+k+1})^2 = \sum_{k=1}^{n} (|z_k - z_{k+1}| + |z_{n+k} - z_{n+k+1}|)^2$$

$$\geq \sum_{k=1}^{n} |z_k - z_{k+1} - z_{n+k} + z_{n+k+1}|^2 = \sum_{k=1}^{n-1} |w_{k+1} - w_k|^2 + |w_n + w_1|^2.$$

On the other hand,

$$\sum_{k=1}^{n} B_k B_{k+n}^2 = \sum_{k=1}^{n} \left| \frac{z_k + z_{k+1}}{2} - \frac{z_{n+k} + z_{n+k+1}}{2} \right|^2$$

$$= \frac{1}{4} \sum_{k=1}^{n-1} |w_k + w_{k+1}|^2 + \frac{1}{4} |w_n - w_1|^2.$$

Hence it is enough to prove the inequality

$$\sum_{k=1}^{n-1} |w_{k+1} - w_k|^2 + |w_n + w_1|^2$$

$$\geq \tan^2 \frac{\pi}{2n} \left(\sum_{k=1}^{n-1} |w_k + w_{k+1}|^2 + \frac{1}{4} |w_n - w_1|^2 \right).$$

Note that this inequality becomes an identity for $n = 2$ and we next assume that $n \geq 3$. Set $w_k = x_k + iy_k$, $x_k, y_k \in \mathbb{R}$, $1 \leq k \leq n$. Then a simple calculation shows that the above inequality can be written as

$$\cos \frac{\pi}{n} \sum_{k=1}^{n} (x_k^2 + y_k^2) \geq \sum_{k=1}^{n-1} (x_k x_{k+1} + y_k y_{k+1}) - x_n x_1 - y_n y_1,$$

which is a consequence of the following inequality:

$$\cos \frac{\pi}{n} \sum_{k=1}^{n} x_k^2 \geq \sum_{k=1}^{n-1} x_k x_{k+1} - x_n x_1,$$

where $n \geq 3$ and $x_1, x_2, \ldots, x_n$ are arbitrary real numbers. This inequality in turn is a consequence of the identity

$$(1) \quad \cos \frac{\pi}{n} \sum_{k=1}^{n} x_k^2 - \sum_{k=1}^{n-1} x_k x_{k+1} + x_n x_1$$

$$= \sum_{k=1}^{n-2} \frac{1}{2 \sin \frac{k\pi}{n} \sin \frac{(k+1)\pi}{n}} \left(\sin \frac{(k+1)\pi}{n} x_k - \sin \frac{k\pi}{n} x_{k+1} + \sin \frac{\pi}{n} x_n \right)^2,$$

which can be proved by comparing the coefficients of x_k^2 and $x_k x_{k+1}$ in both sides of (1). For example, the coefficients of x_n^2 in both sides of (1) coincide because

$$\sum_{k=1}^{n-2} \frac{\sin^2 \frac{\pi}{n}}{2 \sin \frac{k\pi}{n} \sin \frac{(k+1)\pi}{n}} = \sum_{k=1}^{n-2} \frac{\sin \frac{\pi}{n}}{2} \left(\cot \frac{k\pi}{n} - \cot \frac{(k+1)\pi}{n} \right)$$

$$= \frac{\sin \frac{\pi}{n}}{2} \left(\cot \frac{\pi}{n} - \cot \frac{(n-1)\pi}{n} \right) = \cos \frac{\pi}{n}.$$

Remark. Let us discuss the equality case in the given inequality. The above proof shows that for $n = 2$ it is attained only for parallelograms. If $n \geq 3$ the equality is attained if and only if the "opposite" sides of the $2n$-gon $A_1 A_2 \dots A_{2n}$ are parallel and its main diagonals are subject to the following relations for $2 \le k \le n - 1$:

$$(2) \qquad \overrightarrow{A_k A_{n+k}} = \frac{\sin \frac{k\pi}{n}}{\sin \frac{\pi}{n}} \overrightarrow{A_1 A_{n+1}} + \frac{\sin \frac{(k-1)\pi}{n}}{\sin \frac{\pi}{n}} \overrightarrow{A_n A_{2n}}.$$

In particular, we obtain the following generalization of Problem 3 from International Mathematical Olympiad 2003:

Any convex hexagon $A_1 A_2 A_3 A_4 A_5 A_6$ for which $(A_1 A_2 + A_4 A_5)^2 + (A_2 A_3 + A_5 A_6)^2 + (A_3 A_4 + A_6 A_1)^2 = \frac{4}{3}(B_1 B_4^2 + B_2 B_5^2 + B_3 B_6^2)$ is obtained from a triangle by cutting congruent triangles from its "corners" by means of lines parallel to their opposite sides (Fig. 202).

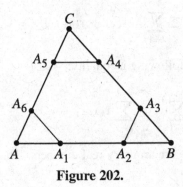

Figure 202.

3.2.11 We use the following lemma.

Lemma. *Let ω be a circle of radius ρ and PR, QS two chords intersecting at X, so that $\angle PXQ = \angle RXS = 2\alpha$. Then $PQ + RS = 4\alpha\rho$ (see Fig. 203).*

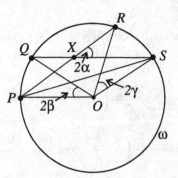

Figure 203.

Proof. Let O be the center of ω and $\angle POQ = 2\beta$, $\angle ROS = 2\gamma$. Then $\angle QSP = \beta$ and $\angle RPS = \gamma$, since the angle at the center is twice the angle at the circumference. Hence $\angle RXS = 2\alpha = \beta + \gamma$ and $PQ + RS = 2\beta\rho + 2\gamma\rho = 4\alpha\rho$. Surround all the given circles with a large circle ω of radius ρ. Consider two circles C_i, C_j with centers O_i, O_j respectively. From the given condition, C_i and C_j do not intersect. Let 2α be the angle between their two internal common tangents PR, QS (Fig. 204).

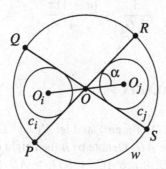

Figure 204.

We have $O_iO_j = 2 \csc \alpha$, so $\alpha \geq \sin \alpha = \dfrac{2}{O_iO_j}$.

Now, from the lemma, $PQ + RS = 4\alpha\rho \geq \dfrac{8\rho}{O_iO_j}$, so that

$$\frac{1}{O_iO_j} \leq \frac{PQ + RS}{8\rho}.$$

We now wish to consider the sum of all these arc lengths as i, j range over all pairs, and we claim that any point of ω is covered by such arcs at most $(n - 1)$ times. To see this, let T be any point of ω and TU a half-line tangent to ω, as in Fig. 205.

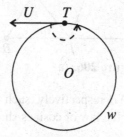

Figure 205.

Consider this half-line as it is rotated about T as shown. At some stage it will intersect a pair of circles for the first time. Relabel these circles C_1 and C_2. The half-line can never intersect three circles, so at some further stage intersection with one of these circles, say C_1, is lost and the half-line will never meet C_1 again during its transit. Continuing in this way and relabeling the circles conveniently, the maximum number of times the half-line can intersect pairs of circles is $(n-1)$, namely when it intersects C_1 and C_2, C_2 and $C_3, \ldots, C_{n-1}$ and C_n. Since T was arbitrary, it follows that the sum of all the arc lengths is less than or equal to $2(n-1)\pi\rho$, and hence

$$\sum_{1 \le i < j \le n} \frac{1}{O_i O_j} \le \frac{(n-1)\pi}{4}.$$

4.12 MaxMin and MinMax

3.3.1 Consider a trapezoid $ABCD$ of area 1 and let C_1 and D_1 be the orthogonal projections of C and D on the line AB. Denote by h the height of $ABCD$. Suppose that $AC_1 \ge BD_1$, i.e., $AC \ge BD$. Since $AC_1 + BD_1 \ge AB + CD$ it follows that $AC_1 \ge \frac{AB+CD}{2}$. Hence $AC_1 \ge \frac{[ABCD]}{h} = \frac{1}{h}$ and we get that

$$AC^2 = AC_1^2 + h^2 \ge \frac{1}{h^2} + h^2 \ge 2.$$

This shows that the least possible length of AC is $\sqrt{2}$.

3.3.2 We first find the minimum side length of an equilateral triangle inscribed in ABC. Let D be a point on BC and put $x = BD$ (Fig. 206 (a)).

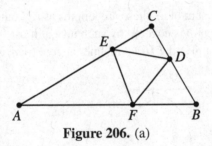

Figure 206. (a)

Then take points E, F on CA, AB respectively, such that $CE = \sqrt{3}x/2$ and $BF = 1 - x/2$. A calculation using the law of cosines shows that

$$DF^2 = DE^2 = EF^2 = \frac{7}{4}x^2 - 2x + 1 = \frac{7}{4}\left(x - \frac{4}{7}\right)^2 + \frac{3}{7}.$$

Hence the triangle DEF is equilateral, and its minimum possible side length is $\sqrt{3/7}$.

We now argue that the minimum possible longest side must occur for some equilateral triangle. Starting with an arbitrary triangle, first suppose it is not isosceles. Then we can slide one of the endpoints of the longest side so as to decrease its length; we do so until there are two longest sides, say DE and EF (Fig. 206 (b)).

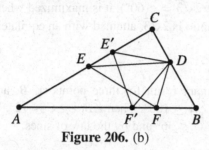

Figure 206. (b)

We now fix D, move E so as to decrease DE, and move F at the same time so as to decrease EF; we do so until all three sides become equal in length. (It is fine if the vertices move onto the extensions of the sides, since the bound above applies in that case as well.)

Hence the minimum is indeed $\sqrt{3/7}$, as desired.

3.3.3 Let the sides of the triangle have lengths $a \leq b \leq c$; let the angles opposite them be A, B, C; let the semiperimeter be $s = \frac{1}{2}(a + b + c)$; and let the inradius be r. Without loss of generality, assume that the triangle has circumradius $R = \frac{1}{2}$. Then the law of sines gives $a = \sin A$, $b = \sin B$, $c = \sin C$.

The area of the triangle equals both $rs = \frac{1}{2}r(\sin A + \sin B + \sin C)$ and $abc/4R = \frac{1}{2}\sin A \sin B \sin C$. Thus

$$r = \frac{\sin A \sin B \sin C}{\sin A + \sin B + \sin C} \quad \text{and} \quad \frac{a}{r} = \frac{\sin A + \sin B + \sin C}{\sin B \sin C}.$$

Because $A = 180° - B - C$, $\sin A = \sin(B + C) = \sin B \cos C + \sin C \cos B$ and we also have

$$\frac{a}{r} = \cot B + \csc B + \cot C + \csc C.$$

Note that the function $f(x) = \cot x + \csc x$ is decreasing along the interval $(0°, 90°)$ since $f'(x) = -\frac{1+\cos x}{\sin^2 x}$.

If $B > 60°$, then $C > B > 60°$ and the triangle with $A' = B' = C' = 60°$ has a larger ratio a'/r'. Therefore we may assume that $B \leq 60°$.

We may further assume that $A = B$; otherwise, the triangle with angles $A' = B' = \frac{1}{2}(A + B) \leq B$ and $C' = C$ has a larger ratio a'/r'. Because $C < 90°$ we have $45° < A \leq 60°$. Now

$$\frac{a}{r} = \frac{\sin A + \sin B + \sin C}{\sin B \sin C} = \frac{2\sin A + \sin(2A)}{\sin A \sin(2A)} = 2\csc(2A) + \csc A.$$

Note that $\csc x$ has second derivative $\csc x (\csc^2 x + \cot^2 x)$, which is strictly positive when $0° < x < 180°$. Thus, both $\csc x$ and $\csc(2x)$ are strictly convex along the interval $0° < x < 90°$. Therefore, $g(A) = 2\csc(2A) + \csc A$, a convex function in A, is maximized in the interval $45° < A \leq 60°$ at one of the endpoints. Because $g(45°) = 2 + \sqrt{2} < 2\sqrt{3} = g(60°)$, it is maximized when $A = 60°$.

Therefore the maximum ratio is $2\sqrt{3}$, attained with an equilateral triangle.

3.3.4

(a) We first prove a preliminary result for three points A, B, and C, under the assumption that $108° \leq \angle A \leq 180°$. Then $\angle B + \angle C \leq 72°$. We may assume that $\angle B \geq \angle C$. Hence $\angle C \leq 36°$ and by the law of sines,

$$\lambda = \frac{BC}{AB} = \frac{\sin(B+C)}{\sin C} \geq \frac{\sin 2C}{\sin C} = 2\cos C \geq 2\cos 36° = 2\sin 54°.$$

Equality holds if and only if $\angle A = 108°$ and $\angle B = \angle C = 36°$.

Consider now any five points in the plane. It follows from our earlier result that $\lambda > 2\sin 54°$ if any three of them are collinear. Henceforth, we assume that this is not the case. Consider the convex hull of the five points. If it is a triangle or a quadrilateral, then one of the five points P is inside the triangle determined by three of the other points. If we join P to these three, the triangle is divided into three smaller triangles. Since the three angles at P sum to $360°$, one of them is at least $120°$. By our earlier result, $\lambda > 2\sin 54°$. If the convex hull is a pentagon, then one of its interior angles is at least $108°$ since the five of them sum to $540°$. Applying our earlier result to the triangle determined by the vertex of this angle and two vertices of the pentagon adjacent to it, we have $\lambda \geq 2\sin 54°$.

(b) From (a), equality can hold only if the convex hull of the five points is a pentagon in which the triangle determined by three adjacent vertices is a $(108°, 36°, 36°)$ triangle. This implies that the pentagon is equilateral, as well as equiangular, so that it is regular. It is easy to verify that for the regular pentagon, we do have $\lambda = 2\sin 54°$.

3.3.5 For a point $P \in C$ denote by P' its antipodal point; for a set $A \subset C$ denote by A' the antipodal image of A (i.e., $A' = \{P' : P \in A\}$).

Take a set $A = \{P_1, P_2, \ldots, P_n\} \subset \mathcal{F}_n$. The set $A \cup A'$ consists of $2m$ points, $m \leq n$, that cut the circle into $2m$ arcs, antipodal in pairs. Denote the set of all these arcs by $\mathcal{A}$.

Let $d = RR'$ be any diameter of C. If $R \in A \cup A'$, then of course $\min_i \delta(P_i, d) = 0$; this trivial situation will be ignored in the sequel. So let us assume that $R \notin A \cup A'$; then R belongs to exactly one arc $\alpha \in A$. The minimum $\min_i \delta(P_i, d)$ occurs when P_i is an endpoint of α (or α') and we get the estimate

$$(1) \qquad\qquad \min_i \delta(P_i, d) \le \sin \frac{\alpha}{2};$$

there should be no ambiguity in denoting the arc and its angular size (length, in other words) by the same symbol α. Equality holds in (1) if and only if R is the midpoint of α.

We seek a diameter d for which the left side of (1) is maximized. This is the case if and only if R is the midpoint of the longest arc in A (there may be more than one pair of such arcs). Denoting by β (the size of) the longest arc in A, we infer that $D(A) = \sin \frac{\beta}{2}$.

Now we wish to minimize this quantity by a suitable choice of A. From among all the $2m$ arcs in A, the longest one has size at least $\frac{\pi}{m}$. Hence

$$(2) \qquad\qquad D(A) \ge \sin \frac{\pi}{2m} \ge \sin \frac{\pi}{2n}.$$

The first inequality in (2) becomes an equality if and only if all arcs in A are equal, i.e., when $A \cup A'$ is the set of vertices of a regular $2m$-gon. The second inequality in (2) is an equality for $m = n$, i.e., when A and A' are disjoint. Hence

$$D_n = \min_{A \in \mathcal{F}_n} D(A) = \sin \frac{\pi}{2n}.$$

The minimum is attained for every set A of n nonantipodal vertices of a regular $2n$-gon inscribed in C.

4.13 Area and Perimeter

3.4.1 Let P be closer to A than to B. Drop the perpendiculars RK and CH onto AB (Fig. 207). Let $AB + BC + CA = 6$.

Then $PQ = AR + AP = 2$ and $AC < AB < 3$. We have

$$AP \le AP + BQ = AB - PQ < 1.$$

Now

$$\frac{[PQR]}{[ABC]} = \frac{PQ \cdot RK}{AB \cdot CH} = \frac{PQ}{AB} \cdot \frac{AR}{AC}.$$

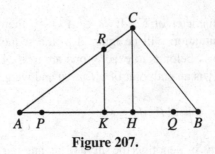

Figure 207.

We have

$$\frac{PQ}{AB} > \frac{2}{3}$$

and

$$\frac{AR}{AC} > \frac{2 - AP}{3} > \frac{1}{3}.$$

It follows that $\frac{[PQR]}{[ABC]} > \frac{2}{9}$.

3.4.2 Let a, b, c be the lengths of the sides opposite angles A, B, C, respectively. By the law of sines,

$$\frac{a}{b} = \frac{\sin 2B}{\sin B} = 2\cos B,$$

$$\frac{c}{b} = \frac{\sin(\pi - 3B)}{\sin B} = \frac{\sin 3B}{\sin B}$$

$$= \frac{(2\sin B \cos B)\cos B + (2\cos^2 B - 1)\sin B}{\sin B} = 4\cos^2 B - 1.$$

Hence $\frac{c}{b} = \left(\frac{a}{b}\right)^2 - 1$, from which

$$(1) \qquad\qquad a^2 = b(b + c).$$

Since we are looking for a triangle of smallest perimeter, we may assume that a, b, and c have no common prime factor; otherwise, a smaller example would exist. In fact, b and c must be relatively prime, for (1) shows that any common prime factor of b and c would be a factor of a as well. Since (1) expresses a perfect square a^2 as the product of two relatively prime integers b and $b + c$, it must be the case that b and $b+c$ are perfect squares. Thus, for some relatively prime integers m and n, we have $b = m^2$, $b + c = n^2$, $a = mn$, and $\frac{n}{m} = \frac{a}{b} = 2\cos B$. The angle $C = \pi - 3B$ is obtuse, so $0 < B < \frac{\pi}{6}$, which implies $\frac{\sqrt{3}}{2} < \cos B < 1$ and thus

$$\sqrt{3} < \frac{n}{m} < 2.$$

It is easy to see that this inequality has no integer solution with $m = 1, 2,$ or 3. Hence $m \geq 4, n \geq 7$, and

$$a + b + c = mn + n^2 \geq 4 \cdot 7 + 7^2 = 77.$$

In fact, the pair $(m, n) = (4, 7)$ generates a triangle with $(a, b, c) = (28, 16, 33)$, and this triangle meets all the necessary geometric conditions, so 77 is the minimum possible perimeter.

3.4.3 Suppose first that two vertices A and B of a triangle ABC lie on the same side PQ of the parallelogram (Fig. 208). Then $AB \leq PQ$, and since the height of $\triangle ABC$ through C is not greater than the height of the parallelogram to PQ, we conclude that the area of $\triangle ABC$ is not greater than one-half the area of the parallelogram.

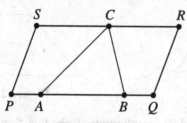

Figure 208.

Assume now that the vertices of the triangle lie on different sides of the parallelogram. Then two of them lie on opposite sides. Draw a line through the third vertex of the triangle that is parallel to these sides. It divides the parallelogram into two parallelograms and the triangle into two triangles (Fig. 209), and we can apply the same reasoning as in the first case.

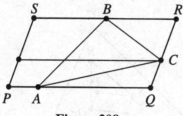

Figure 209.

3.4.4 Suppose first that the parallelogram $EFGH$ is inscribed in triangle ABC so that $E, F \in AB, G \in BC$, and $H \in CA$ (Fig. 210).

Set $CH : CA = x$, where $0 < x < 1$. Then it is easy to show that $S = 2x(1-x)T$ and the arithmetic mean–geometric mean inequality gives $2S \leq T$.

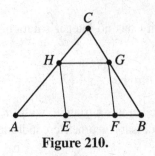

Figure 210.

In the general case draw parallel lines containing two opposite sides of the parallelogram. If these two lines intersect only two sides of the triangle, then the problem can be reduced to the case considered above (Fig. 211).

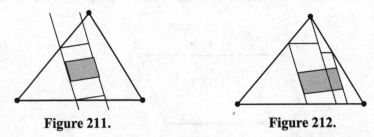

Figure 211.   **Figure 212.**

Otherwise it can be reduced to a configuration like that shown in Fig. 212 and one applies again the case considered above.

3.4.5 Let the angles at P, Q, R, and S be α, β, γ, and δ, respectively. Since $(\alpha + \beta) + (\gamma + \delta) = 360°$, we may assume that $\alpha + \beta \geq 180°$.

Similarly, we may assume that $\alpha + \delta \geq 180°$. Complete the parallelogram $PQTS$ (Fig. 213). Then T must lie inside $PQRS$, and hence inside ABC. Now $[PQS] = \frac{1}{2}[PQTS] \leq \frac{1}{4}[ABC]$ by Problem 3.4.4.

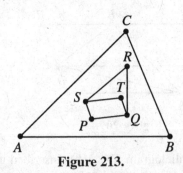

Figure 213.

3.4.6 Let M be a polygon with center of symmetry O contained in a triangle ABC. For any point X in the plane denote by X' the symmetric point of X with respect to

O. Then M is contained in the common part T of triangles ABC and $A'B'C'$. Note that O is the center of symmetry of the polygon T. Since $AB \| A'B'$, $BC \| B'C'$, $CA \| C'A'$ and $AB = A'B'$, $BC = B'C'$, $CA = C'A'$ it follows that at least two vertices of $\triangle A'B'C'$ lie outside $\triangle ABC$. Suppose first that A' lies in the interior of $\triangle ABC$. Then T is a parallelogram, and it follows from Problem 3.4.4 that $[M] \leq [T] \leq \frac{1}{2}[ABC]$. Let now the points A', B', and C' lie outside $\triangle ABC$. Then T is a hexagon $A_1 A_2 B_1 B_2 C_1 C_2$ as shown in Fig. 214.

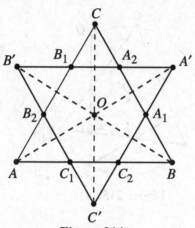

Figure 214.

Set $\frac{AC_1}{AB} = \frac{AB_2}{AC} = x$, $\frac{BC_2}{AB} = \frac{BA_1}{BC} = y$, $\frac{CA_2}{CA} = \frac{CB_1}{CB} = z$. Note that C_1' lies on the lines $A'B'$ and BC, i.e., $C_1' = A_2$. Similarly, $C_2' = B_1$ and therefore $C_1 C_2 = B_1 A_2$. Hence $\frac{C_1 C_2}{AB} = \frac{B_1 A_2}{AB} = z$ and we get

$$x + y + z = \frac{AC_1}{AB} + \frac{BC_2}{AB} + \frac{C_1 C_2}{AB} = 1.$$

On the other hand,

$$[T] = [ABC] - [AC_1 B_2] - [BA_1 C_2] - [CB_1 A_2] = [ABC](1 - x^2 - y^2 - z^2).$$

Now the root mean square–arithmetic mean inequality gives

$$x^2 + y^2 + z^2 \geq \frac{1}{3}(x + y + z)^2 = \frac{1}{3}$$

and we get $[T] \leq \frac{2}{3}[ABC]$. Equality holds if and only if $x = y = z = \frac{1}{3}$, i.e., when the points A_1 and A_2, B_1 and B_2, C_1 and C_2 divide the sides BC, CA, and AB into three equal parts. Thus the solution of the problem is given by the hexagon $A_1 A_2 B_1 B_2 C_1 C_2$.

3.4.7 Denote the triangles by ABC and PQR, and let D and E be the points of intersection of AB with PR and AB with PQ, respectively (Fig. 215). Then by rotational symmetry, the entire figure is symmetric about the line OD, and also the line OE, where O is the center of the circle. Moreover,

$$K = [ABC] - 3[PDE].$$

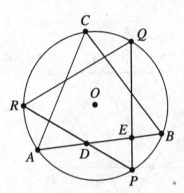

Figure 215.

So K will be a minimum when $\triangle PDE$ has maximum area. Note that $PD = AD$, $PE = BE$, so that $\triangle PDE$ has the constant perimeter $AB = r\sqrt{3}$. It follows from Problem 1.2.1 that $\triangle PDE$ has maximum area when P is the midpoint of arc AB. In this case the sides of $\triangle PDE$ are $\frac{1}{3}$ as long as the sides of $\triangle ABC$, so $[PDE] = \frac{1}{9}[ABC]$. Hence

$$K \geq [ABC]\left(1 - \frac{3}{9}\right) = \frac{2}{3}(r\sqrt{3})^2 \frac{\sqrt{3}}{4} = \frac{\sqrt{3}r^2}{2}.$$

Remark. In a similar fashion, one can obtain the analogous area inequality for two regular n-gons inscribed in a circle. K will be minimum when one of the n-gons can be obtained from the other one by rotation of $\frac{\pi}{n}$ about the center.

3.4.8 It is clear that if a triangle contains another triangle then the inradius of the first one is not less than the inradius of the second one. This remark easily leads to the conclusion that it is enough to consider only triangles ABC like the one shown in Fig. 216.

Denote by r the inradius of $\triangle ABC$. Set $PC = a$, $BM = b$ and let N be the point on the ray PC such that $PN = a + b$. Set $x = AC = \sqrt{1 + a^2}$, $y = BC = \sqrt{(1 - a)^2 + (1 - b)^2}$, $z = AB = \sqrt{1 + b^2}$, $u = AN = \sqrt{1 + (a + b)^2}$, $v = NM = \sqrt{1 + (1 - a - b)^2}$. Then $u \geq z \geq 1$ and $x \geq 1$, $v \geq 1$ and we get

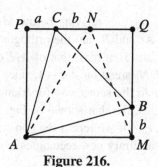

Figure 216.

$$(u + v + 1) - (x + y + z) = \frac{u^2 - x^2}{u + x} + \frac{v^2 - y^2}{v + y} + \frac{1^2 - z^2}{1 + z}$$

$$= \frac{2ab + b^2}{u + x} + \frac{2ab}{v + y} - \frac{b^2}{1 + z} \le \frac{2ab + b^2}{1 + z} + \frac{2ab}{v + y} - \frac{b^2}{1 + z}$$

$$= 2ab \left(\frac{1}{v + y} + \frac{1}{1 + z} \right) \le 2ab \left(\frac{1}{2} + 1 \right) = 3ab \le (u + v + 1)ab.$$

Hence

$$\frac{1 - ab}{x + y + z} \le \frac{1}{u + v + 1}.$$

On the other hand,

$$[ABC] = 1 - \frac{a}{2} - \frac{b}{2} - \frac{(1 - a)(1 - b)}{2} = \frac{1 - ab}{2}$$

and therefore

$$r = \frac{2[ABC]}{x + y + z} = \frac{1 - ab}{x + y + z} \le \frac{1}{u + v + 1}.$$

Now using Heron's problem (Problem 1.1.1) we see that $u + v = AN + MN$ is a minimum if N is the midpoint of PQ, i.e.,

$$r \le \frac{1}{u + v + 1} \le \frac{1}{\sqrt{5} + 1} = \frac{\sqrt{5} - 1}{4}.$$

Thus the maximum value of r is equal to $\frac{\sqrt{5}-1}{4}$ and it is attained only if $B = M$ and C is the midpoint of PQ.

3.4.9 We show first that the rectangles must be placed one over another as shown in Fig. 217. Indeed, let $r_1, r_2, \ldots, r_n$ be arbitrary nonintersecting rectangles in $\triangle ABC$ with a side parallel to AB. Consider the lines determined by their upper sides and let the one closest to AB intersect AC and BC at points M_1 and N_1, respectively. Then the parts of $r_1, r_2, \ldots, r_n$ lying below $M_1 N_1$ are contained in the rectangle $A_1 B_1 N_1 M_1$, where A_1 and B_1 are the orthogonal projections of M_1 and N_1 on AB. Hence their total area is at most that of $A_1 B_1 M_1 N_1$. Note that the parts $r_1, r_2, \ldots, r_n$ lying above $M_1 N_1$ are at most $n - 1$, since $A_1 B_1 M_1 N_1$ contains at least one of them. Proceeding in the same way for triangle $M_1 N_1 C$, etc., we conclude that there are n rectangles like that shown in Fig. 217 whose total area is not less than that of $r_1, r_2, \ldots, r_n$. (If we repeat the construction above k times where $k < n$, then we add $n - k$ arbitrary new rectangles constructed in the same way in triangle $M_k N_k C$.)

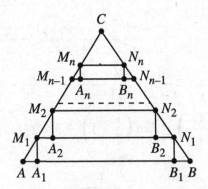

Figure 217.

Denote by x_k, $1 \le k \le n$, the distance between the parallel lines $M_k N_k$ and $M_{k-1} N_{k-1}$ ($M_0 = A$, $N_0 = B$) and by x_{n+1} the distance from C to $M_n N_n$ (Fig. 217). Let CC_0 be the altitude of $\triangle ABC$ through C and $h = CC_0$. Then $\triangle M_{k-1} A_k M_k \sim \triangle A C_0 C$ and we get $[M_{k-1} A_k M_k] = \frac{x_k^2}{h^2}[AC_0 C]$. Likewise $[N_{k-1} B_k N_k] = \frac{x_k^2}{h^2}[BC_0 C]$, $[M_n N_n C] = \frac{x_{n+1}^2}{h^2}[ABC]$. Denote by S_n the combined area of rectangles $A_k B_k N_k M_k$, $1 \le k \le n$. Then

$$S_n = [ABC] - [M_n N_n C] - \sum_{k=1}^{n}([M_{k-1} A_k M_k] + [N_{k-1} B_k N_k])$$

$$= [ABC]\left(1 - \frac{1}{h^2}\sum_{k=1}^{n+1} x_k^2\right).$$

Taking into account that $\sum_{k=1}^{n+1} x_k = h$ we get from root mean square–arithmetic mean inequality that

$$\sum_{k=1}^{n+1} x_k^2 \ge \frac{1}{n+1}\left(\sum_{k=1}^{n+1} x_k\right)^2 = \frac{h^2}{n+1}.$$

Hence $S_n \leq \frac{n}{n+1}[ABC]$, with equality only if $x_1 = x_2 = \cdots = x_{n+1} = \frac{h}{n+1}$. Thus, the desired rectangles must be cut as in Fig. 217, where the points $M_1, M_2, \ldots, M_n$ $(N_1, N_2, \ldots, N_n)$ divide $AC(BC)$ into $n + 1$ equal parts.

3.4.10 We deduce from the area of $P_1 P_3 P_5 P_7$ that the radius of the circle is $\sqrt{5/2}$. An easy calculation using the Pythagorean theorem then shows that the rectangle $P_2 P_4 P_6 P_8$ has sides $\sqrt{2}$ and $2\sqrt{2}$.

By symmetry, the area of the octagon can be expressed as (Fig. 218)

$$[P_2 P_4 P_6 P_8] + 2[P_2 P_3 P_4] + 2[P_4 P_5 P_6].$$

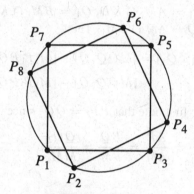

Figure 218.

Note that $[P_2 P_3 P_4]$ is $\sqrt{2}$ times the distance from P_3 to $P_2 P_4$, which is maximized when P_3 lies on the midpoint of arc $P_2 P_4$; similarly, $[P_4 P_5 P_6]$ is $2\sqrt{2}$ times the distance from P_5 to $P_4 P_6$, which is maximized when P_5 lies on the midpoint of arc $P_4 P_6$.

Thus, the area of the octagon is maximized when P_3 is the midpoint of arc $P_2 P_4$ and P_5 is the midpoint of arc $P_4 P_6$. In this case, it is easy to calculate that $[P_2 P_3 P_4] = \sqrt{5} - 1$ and $[P_4 P_5 P_6] = \sqrt{5}/2 - 1$ and so the area of the octagon is $3\sqrt{5}$.

3.4.11 We shall show that the desired point M is such that $\frac{DM}{MC} = \frac{AK}{KB}$.

Let P and Q be the intersection points of AM and DK, and BM and CK, respectively (Fig. 219). Then

$$\frac{KQ}{QC} = \frac{KB}{MC} = \frac{AK}{DM} = \frac{KP}{PD},$$

which shows that $PQ \| CD \| AB$.

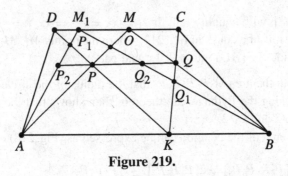

Figure 219.

Now consider an arbitrary point $M_1 \neq M$ on DC. We may assume that M_1 lies between D and M. Set $P_1 = AM_1 \cap KD$, $Q_1 = BM_1 \cap KC$, $P_2 = AM_1 \cap PQ$, $Q_2 = BM_1 \cap PQ$, and $O = AM \cap BM_1$. Then

$$[MPKQ] - [M_1P_1KQ_1] = [MOQ_1Q] - [M_1P_1PO]$$
$$> [MOQ_2Q] - [M_1P_2PO] = 0.$$

To prove the latter equality we first note that $PP_2 = QQ_2$ since

$$\frac{PP_2}{MM_1} = \frac{AP}{AM} = \frac{BQ}{BM} = \frac{QQ_2}{MM_1}.$$

Hence

$$[MOQ_2Q] = [MPQ] - [OPQ_2] = [M_1P_2Q_2] - [OPQ_2] = [M_1P_2PO].$$

3.4.12 *First solution.* Let a, b, c denote the lengths of the sides BC, CA, AB, respectively. We assume without loss of generality that $a \leq b \leq c$.

Choose l to be the angle bisector of $\angle A$. Let P be the intersection point of l with BC (Fig. 220). Since $AC \leq AB$, the intersection of triangles ABC and $A'B'C'$ is the disjoint union of two congruent triangles, APC and APC'.

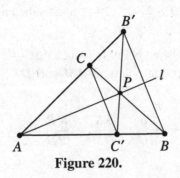

Figure 220.

Considering BC as a base, triangles APC and ABC have equal altitudes, so their areas are in the same ratio as their bases:

$$\frac{[APC]}{[ABC]} = \frac{PC}{BC}.$$

Since AP is the angle bisector of $\angle A$, we have $BP/PC = c/b$, so

$$\frac{PC}{BC} = \frac{PC}{BP + PC} = \frac{b}{b+c}.$$

But $2b \geq a + b > c$ by the triangle inequality and we get

$$\frac{[AC'PC]}{[ABC]} = \frac{2[APC]}{[ABC]} = \frac{2b}{b+c} > \frac{2}{3}.$$

Second solution. Let the foot of the altitude from C meet AB at D.

First suppose $[BDC] > \frac{1}{3}[ABC]$. In this case we reflect through CD. If B' is the image of B, then $BB'C$ lies in ABC and the area of the overlap is at least $\frac{2}{3}[ABC]$.

Now suppose $[BDC] \leq \frac{1}{3}[ABC]$. In this case we reflect through the bisector of $\angle A$. If C' is the image of C, then triangle ACC' is contained in the overlap, and $[ACC'] > [ADC] \geq \frac{2}{3}[ABC]$.

Remark. Let F denote the figure given by the intersection of the interior of triangle ABC and the interior of its reflection in l. Yet another approach to the problem involves finding the maximum attained for $[F]/[ABC]$ by taking l from the family of lines perpendicular to AB. By choosing the best alternative between the angle bisector at C and the optimal line perpendicular to AB, one can ensure

$$\frac{[F]}{[ABC]} > \frac{2}{1 + \sqrt{2}} = 2(\sqrt{2} - 1) = 0.828427\ldots,$$

and this constant is in fact the best possible.

3.4.13 The key observation is that for any side S of P_6, there is some subsegment of S that is a side of P_n. (This is easily proved by induction on n.) Thus P_n has a vertex on each side of P_6. Since P_n is convex, it contains a hexagon Q with (at least) one vertex on each side of P_6. (The hexagon may be degenerate, since some of its vertices may coincide.)

Let $P_6 = A_1 A_2 A_3 A_4 A_5 A_6$ and let $Q = B_1 B_2 B_3 B_4 B_5 B_6$, with B_i on $A_i A_{i+1}$ (indices are considered modulo 6).

The side $B_i B_{i+1}$ of Q is entirely contained in triangle $A_i A_{i+1} A_{i+2}$, so Q encloses the smaller regular hexagon R (shaded in Fig. 221) whose sides are the central thirds of the segments $A_i A_{i+2}$, $1 \leq i \leq 6$. The area of R is $1/3$, as can be

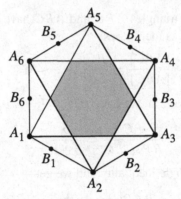

Figure 221.

seen from the fact that its side length is $1/\sqrt{3}$ times the side length of P_6. Thus $[P_n] \geq [Q] \geq [R] = 1/3$. We obtain strict inequality by observing that P_n is strictly larger than Q: if $n = 6$, this is obvious; if $n > 6$, then P_n cannot equal Q because P_n has more sides.

Remark. With a little more work, one could improve $1/3$ to $1/2$. The minimum area of a hexagon Q with one vertex on each side of P_6 is in fact $1/2$, attained when the vertices of Q coincide in pairs at every other vertex of P_6. So, the hexagon Q degenerates into an equilateral triangle. This can be done using the same arguments as those in the solution of Problem 1.4.2. If the conditions of the problem were changed so that the cut-points could be anywhere within adjacent segments instead of just at the midpoints, then the best possible bound would be $1/2$.

3.4.14 Note first that the area of any triangle whose vertices have integer coordinates is a number of the form $\frac{n}{2}$, where n is a positive integer. To prove this consider the smallest rectangle containing the triangle and whose vertices have integer coordinates (Fig. 222). Hence the area of any such triangle is at least $\frac{1}{2}$. Thus

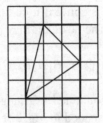

Figure 222.

it is enough to prove that the given pentagon $A_1A_2A_3A_4A_5$ contains a point with integer coordinates in its interior.

Assume the contrary. It is easy to see that there exist two vertices of the pentagon, say $A_i(x_i, y_i)$ and $A_j(x_j, y_j)$, such that $x_i \equiv x_j \pmod 2$ and $y_i \equiv y_j \pmod 2$. Then the midpoint $M\left(\frac{x_i+x_j}{2}, \frac{y_i+y_j}{2}\right)$ of the segment A_iA_j has integer coordinates and therefore it lies on the boundary of the pentagon. Hence A_i and A_j are consecutive vertices and let $A_i = A_1$, $A_j = A_5$. Applying the same arguments to the pentagon $A_1A_2A_3A_4M$ and so on, we obtain infinitely many points with integer coordinates on the boundary of the pentagon, a contradiction.

3.4.15

Lemma 1 *If* $0 \le x, y \le 1$, *then*

$$\sqrt{1-x^2} + \sqrt{1-y^2} \ge \sqrt{1-(x+y-1)^2}.$$

Proof. Squaring and subtracting $2 - x^2 - y^2$ from both sides gives the equivalent inequality $2\sqrt{(1-x^2)(1-y^2)} \ge -2(1-x)(1-y)$. It is true because the left side is nonnegative and the right is nonpositive.

Lemma 2 *If* $x_1 + \cdots x_n \le n - \frac{1}{2}$ *and* $0 \le x_i \le 1$ *for each i, then*

$$\sum_{i=1}^{n} \sqrt{1-x_i^2} \ge \frac{\sqrt{3}}{2}.$$

Proof. We use induction on n. In the case $n = 1$, the statement is clear. If $n > 1$, then either $\min(x_1, x_2) \le \frac{1}{2}$ or $x_1 + x_2 > 1$. In the first case we immediately have $\max\left(\sqrt{1-x_1^2}, \sqrt{1-x_2^2}\right) \ge \frac{\sqrt{3}}{2}$. In the second case, we can replace x_1 and x_2 by the single number $x_1 + x_2 - 1$ and use the induction hypothesis together with the previous lemma.

Let P and Q be vertices of our polygon such that $l = PQ$ is a maximum. The polygon consists of two paths from P to Q, each of intgral length greater than or equal to l; these lengths are distinct because the perimeter is odd. Then the greater of the two lengths, m, is at least $l + 1$. Position the polygon in the coordinate plane with $P = (0, 0)$, $Q = (l, 0)$ and the longer path in the upper half-plane. Because each side of the polygon has integer length, we can divide this path into segments of length 1. Let the endpoints of these segments, in order, be $P_0 = P$, $P_1 = (x_1, y_1)$, $P_2 = (x_2, y_2), \ldots, P_m = Q$. There exists some r such that y_r is a maximum. Then either $r \ge x_r + \frac{1}{2}$ or $(m - r) \ge (l - x_r) + \frac{1}{2}$. Assume the former (otherwise, just reverse the choices of P and Q). We already know that

$y_1 \geq 0$, and by the maximal definition of l we must have $x_1 \geq 0$ as well. Because the polygon is convex, we must have $y_1 \leq y_2 \leq \cdots \leq y_r$ and $x_1 \leq x_2 \leq \cdots \leq x_r$. Now $y_{i+1} - y_i = \sqrt{1 - (x_{i+1} - x_i)^2}$, so

$$y_r = \sum_{i=0}^{r-1} (y_{i+1} - y_i) = \sum_{i=0}^{r-1} \sqrt{1 - (x_{i+1} - x_i)^2} \geq \frac{\sqrt{3}}{2}$$

by the second lemma. Hence triangle PP_rQ has base PQ with length at least 1 and height $y_r \geq \frac{\sqrt{3}}{2}$, implying that its area is at least $\frac{\sqrt{3}}{4}$. Because our polygon is convex, it contains this triangle, and hence the area of the whole polygon is also at least $\frac{\sqrt{3}}{4}$.

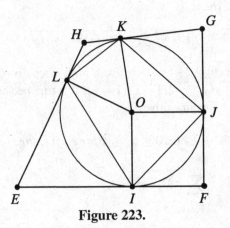

Figure 223.

3.4.16 Let the outer quadrilateral be $EFGH$ with angles $\angle E = 2\alpha_1$, $\angle F = 2\alpha_2$, $\angle G = 2\alpha_3$, $\angle H = 2\alpha_4$. Let the circumcircle of C have radius r and center O, and let the sides EF, FG, GH, HE be tangent to C at I, J, K, L (Fig. 223). In the right triangle EIO, we have $IO = r$ and $\angle OEI = \alpha_1$, so that $EI = r \cot \alpha_1$. After finding similar expressions for $IF, FJ, \ldots, LE$, we have that $P_T = 2r \sum_{i=1}^{4} \cot \alpha_i$. Also, $[EFO] = \frac{1}{2} EF \cdot IO = \frac{1}{2} EF \cdot r$. Finding $[FGO]$, $[GHO]$, $[HEO]$ similarly shows that $A_T = \frac{1}{2} P_T \cdot r$. Note that

$$IJ = 2r \sin \angle IOF = 2r \sin(90° - \alpha_2) = 2r \cos \alpha_2.$$

Similar expressions hold for JK, KL, LI leading to $P_C = 2r \sum_{i=1}^{4} \cos \alpha_i$. Also note that $\angle IOJ = 180° - \angle JFI = 180° - 2\alpha_2$ and hence

$$[IOJ] = \frac{1}{2} OI \cdot OJ \sin \angle IOJ = \frac{1}{2} r^2 \sin 2\alpha_2 = r^2 \sin \alpha_2 \cos \alpha_2.$$

Adding this to the analogous expressions for $[JOK]$, $[KOL]$, $[LOI]$, we find that

$$Ac = r^2 \sum_{i=1}^{4} \sin \alpha_i \cos \alpha_i.$$

Therefore the inequality we wish to prove is equivalent to

$$\left(\sum_{i=1}^{4} \cot \alpha_i \right) \left(\sum_{i=1}^{4} \sin \alpha_i \cos \alpha_i \right) \geq \left(\sum_{i=1}^{4} \cos \alpha_i \right)^2,$$

which is true by the Cauchy–Schwarz inequality.

3.4.17

(a) Let O be the common center of the two circles (Fig. 224). Applying Ptolemy's inequality (Problem 3.2.6) to the quadrilaterals OAB_1C_1, OBC_1D_1, OCD_1A_1, and ODA_1B_1, we have

$$R \cdot AC_1 \leq r \cdot B_1C_1 + R \cdot AB_1,$$

$$R \cdot BD_1 \leq r \cdot C_1D_1 + R \cdot BC_1,$$

$$R \cdot CA_1 \leq r \cdot D_1A_1 + R \cdot CD_1,$$

$$R \cdot DB_1 \leq r \cdot A_1B_1 + R \cdot DA_1.$$

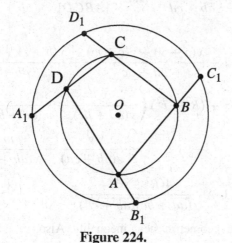

Figure 224.

Addition yields

$$R \cdot (AB + BC + CD + DA) \leq r \cdot (A_1B_1 + B_1C_1 + C_1D_1 + D_1A_1).$$

For equality to hold, all four quadrilaterals must be cyclic. Hence

$$\angle OAC_1 = \angle OB_1C_1 = \angle OC_1B_1 = \angle OAD$$

by Thales' theorem, so that OA bisects $\angle BAD$. Similarly, OB, OC, and OD bisect $\angle ABC$, $\angle BCD$, and $\angle CDA$, respectively. Hence O is also the incenter of $ABCD$. This is possible only if $ABCD$ is a square. Conversely, if $ABCD$ is a square, so is $A_1B_1C_1D_1$, and the perimeter of the latter is clearly $\frac{R}{r}$ times that of the former.

(b) Let $a = AB, b = BC, c = CD, d = DA, w = A_1D, x = B_1A, y = C_1B$, and $z = D_1C$. By the power-of-a-point theorem (Fig. 224),

$$x(x + d) = y(y + a) = z(z + b) = w(w + c) = R^2 - r^2.$$

Since we have

$$\angle B_1AC_1 = 180° - \angle DAB = \angle BCD = 180° - \angle A_1CD_1,$$

we also have

$$\frac{[AB_1C_1]}{[ABCD]} = \frac{x(a + y)}{ad + bc} \quad \text{and} \quad \frac{[A_1CD_1]}{[ABCD]} = \frac{z(c + w)}{ad + bc}.$$

Similarly,

$$\frac{[BC_1D_1]}{[ABCD]} = \frac{y(b + z)}{ab + cd} \quad \text{and} \quad \frac{[A_1B_1D]}{[ABCD]} = \frac{w(d + x)}{ab + cd}.$$

Hence

$$\frac{[A_1B_1C_1D_1]}{[ABCD]} = 1 + \frac{x(a + y) + z(c + w)}{ad + bc} + \frac{y(b + z) + w(d + x)}{ab + cd}$$

$$= 1 + (R^2 - r^2)\left(\frac{x}{y(ad + bc)} + \frac{z}{w(ad + bc)}\right.$$

$$\left. + \frac{y}{z(ab + cd)} + \frac{w}{x(ab + cd)}\right)$$

$$\geq 1 + \frac{4(R^2 - r^2)}{\sqrt{(ad + bc)(ab + cd)}}$$

by the arithmetic mean–geometric mean inequality. Also,

$$2\sqrt{(ad + bc)(ab + cd)} \leq (ad + bc) + (ab + cd)$$

$$= (a + c)(b + d) \leq \frac{1}{4}(a + b + c + d)^2 \leq 8r^2.$$

The last step uses the fact that among all quadrilaterals inscribed in a circle, the square has the greatest perimeter (Problem 2.1.12). We now have

$$\frac{[A_1 B_1 C_1 D_1]}{[ABCD]} \geq 1 + \frac{4(R^2 - r^2)}{4r^2} = \frac{R^2}{r^2}.$$

3.4.18

(a) Let ABC be a right-angled triangle whose vertices are grid points and whose legs go along the lines of the grid with $\angle A = 90°$, $AB = m$, and $AC = n$. Let us consider the $m \times n$ rectangle $ABCD$ as shown in Fig. 225.

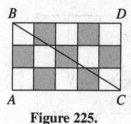

B   D

A   C

Figure 225.

For an arbitrary polygon P let us denote by $S_b(P)$ the total area of the black part of P and by $S_w(P)$ the total area of its white part.

When m and n are of the same parity the coloring of the rectangle $ABCD$ is centrally symmetric about the midpoint of the hypotenuse BC. Hence $S_b(ABC) = S_b(BCD)$ and $S_w(ABC) = S_w(BCD)$. Therefore

$$f(m, n) = |S_b(ABC) - S_w(ABC)| = \frac{1}{2}|S_b(ABCD) - S_w(ABCD)|.$$

Hence $f(m, n) = 0$ for m, n both even and $f(m, n) = \frac{1}{2}$ for m, n both odd.

(b) If m, n are both even or both odd the result follows from (a). Suppose now that m is odd and n is even. Consider a point L on AB such that $AL = m - 1$ as shown in Fig. 226.

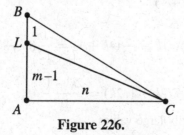

B

1

L

$m-1$

n

A   C

Figure 226.

Since $m - 1$ is even we have $f(m - 1, n) = 0$, i.e., $S_b(ALC) = S_w(ALC)$. Therefore

$$f(m, n) = |S_b(ABC) - S_w(ABC)| = |S_b(LBC) - S_w(LBC)|$$

$$\leq [LBC] = \frac{n}{2} \leq \frac{1}{2} \max(m, n).$$

(c) Let us compute $f(2k + 1, 2k)$. As in (b) we will consider a point L on AB such that $AL = 2k$. Since $f(2k, 2k) = 0$ and $S_b(ALC) = S_w(ALC)$, we have

$$f(2k + 1, 1k) = |S_b(LBC) - S_w(LBC)|.$$

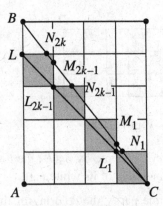

Figure 227.

The area of the triangle LBC is k. Suppose without loss of generality that the diagonal LC is all black (see Fig. 227). Then the white part of LBC consists of several triangles $BLN_{2k}, M_{2k-1}L_{2k-1}N_{2k-1}, M_1L_1N_1$ each of them similar to BAC. Their total area is

$$S_w(LBC) = \frac{1}{2} \frac{2k}{2k + 1} \left(\left(\frac{2k}{2k} \right)^2 + \left(\frac{2k - 1}{2k} \right)^2 + \cdots + \left(\frac{1}{2k} \right)^2 \right)$$

$$= \frac{1}{4k(2k + 1)} (1^2 + 2^2 + \cdots + (2k)^2) = \frac{4k + 1}{12}.$$

Therefore

$$S_b(LBC) = k - \frac{1}{12}(4k + 1) = \frac{1}{12}(8k - 1)$$

and thus

$$f(2k + 1, 2k) = \frac{2k - 1}{6}.$$

This function takes arbitrarily large values.

4.14 Polygons in a Square

3.5.1 *Hint.* Use the same arguments as in the solution of Problem 3.4.3.

3.5.2 Let $ABCD$ be a unit square and $MNPQ$ a quadrilateral inscribed in it (Fig. 228).

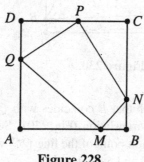

Figure 228.

Suppose that all its sides have lengths less than $\frac{\sqrt{2}}{2}$. Then the root mean square–arithmetic mean inequality implies that

$$AM + AQ \leq \sqrt{2(AM^2 + AQ^2)} = \sqrt{2MQ^2} < 1.$$

Analogously, $MB + MN < 1$, $CN + CP < 1$, and $PD + DQ < 1$. Adding these inequalities gives $4 = AB + BC + CD + DA < 4$, a contradiction.

3.5.3 The side length of any equilateral triangle inscribed in a unit square is at least 1, since two of its vertices lie on opposite sides of the square. Hence the minimum of its area is equal to $\frac{\sqrt{3}}{4}$, and it is attained when one of its sides is parallel to a side of the square (Fig. 229).

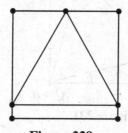

Figure 229.

Let now $PQRS$ be a unit square and ABC an equilateral triangle such that $A \in PS$, $B \in QR$, and $C \in SR$ (Fig. 230). We may assume that $AP \geq BQ$.

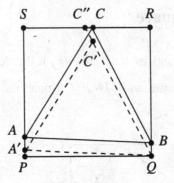

Figure 230.

Translate $\triangle ABC$ vertically such that B coincides with Q and let A' and C' be the images of A and C under this translation (Fig. 230). Set $\alpha = \angle A'QP, \beta = \angle C'QR$ and let C'' be the intersection point of the line QC' with SR. Suppose that $\alpha > 15°$. Then $\beta = 30° - \alpha < 15°$ and we get

$$C'Q = A'Q = \frac{1}{\cos\alpha} > \frac{1}{\cos\beta} = C''Q,$$

a contradiction. Hence $\alpha \le 15°$ and we have $AB = A'Q = \frac{1}{\cos\alpha} \le \frac{1}{\cos 15°}$. This inequality shows that the area of ABC is a maximum when $B = Q$ and $\angle AQP = \angle CQR = 15°$ (Fig. 231). Note that in this case

$$[ABC] = \frac{AB^2\sqrt{3}}{4} = \frac{\sqrt{3}}{4\cos^2 15°} = \frac{\sqrt{3}}{2(1+\cos 30°)} = 2\sqrt{3} - 3.$$

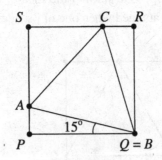

Figure 231.

3.5.4 Draw lines parallel to a side of the square through all vertices of the given polygon. They divide it into triangles and trapezoids (Fig. 232). Consider the line segments joining the midpoints of their sides that are not parallel to the drawn lines. Suppose that all of them have lengths less than $\frac{1}{2}$. Since the total length of the heights of all triangles and trapezoids is less than 1, it follows that the area of the polygon is less than $\frac{1}{2}$, a contradiction.

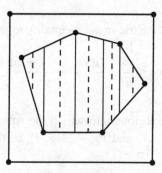

Figure 232.

3.5.5 We shall show that there exist three consecutive vertices of the n-gon having the desired property. Denote by $a_1, a_2, \ldots, a_n$ the side lengths of the n-gon. Let α_i be the angle between its ith and $(i+1)$th sides and let S_i be the area of the triangle formed by these sides. Then $S_i = \frac{1}{2} a_i a_{i+1} \sin \alpha_i$, $1 \le i \le n$.

(a) Denote by S the least of the numbers $S_1, S_2, \ldots, S_n$. Then

$$(2S)^n \le (2S_1)(2S_2) \cdots (2S_n)$$
$$= (a_1 a_2 \cdots a_n)^2 \sin \alpha_1 \sin \alpha_2 \cdots \sin \alpha_n \le (a_1 a_2 \cdots a_n)^2,$$

and the arithmetic mean–geometric mean inequality gives

$$\tag{1} 2S \le (a_1 a_2 \cdots a_n)^{\frac{2}{n}} \le \left(\frac{a_1 + a_2 + \cdots + a_n}{n} \right)^2 .$$

Denote by p_i and q_i the lengths of the orthogonal projections of the ith side of the n-gon on two perpendicular sides of the square. Then $a_i \le p_i + q_i$, $1 \le i \le n$, and we get

$$a_1 + a_2 + \cdots + a_n \le (p_1 + p_2 + \cdots + p_n) + (q_1 + q_2 + \cdots + q_n) \le 4.$$

Thus (1) implies $S \le \frac{8}{n^2}$.

(b) The function $\sin x$ is concave along the interval $[0, \pi]$ since $(\sin x)'' = -\sin x < 0$. Hence Jensen's inequality gives

$$\tag{2} \frac{\sin \alpha_1 + \cdots + \sin \alpha_n}{n} \le \sin \frac{\alpha_1 + \cdots + \alpha_n}{n} = \sin \frac{2\pi}{n} .$$

On the other hand, using the same arguments as in (a) we get

$$2S \leq (a_1 a_2 \cdots a_n)^{\frac{2}{n}} (\sin \alpha_1 \cdots \sin \alpha_n)^{\frac{1}{n}} \leq \frac{16}{n^2} (\sin \alpha_1 \cdots \sin \alpha_n)^{\frac{1}{n}}.$$

Now the arithmetic mean–geometric mean inequality together with (2) gives

$$S \leq \frac{8}{n^2} (\sin \alpha_1 \cdots \sin \alpha_n)^{\frac{1}{n}} \leq \frac{8}{n^2} \left(\frac{\sin \alpha_1 + \cdots + \sin \alpha_n}{n} \right) \leq \frac{8}{n^2} \sin \frac{2\pi}{n}.$$

3.5.6 Let $D_1, D_2, \ldots, D_n$ be the regions into which the square is divided by the line segments and let $A_1, A_2, \ldots, A_n$ and $P_1, P_2, \ldots, P_n$ be their respective areas and perimeters. Then

$$\sum_{i=1}^{n} P_i = 4 + 2 \cdot 18 = 40.$$

Consider an arbitrary region D_i. Let Q_i be the smallest rectangle circumscribed around D_i (Fig. 233).

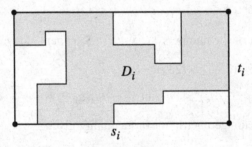

Figure 233.

Clearly $P_i \geq 2(s_i + t_i)$ and $A_i \leq s_i t_i$. Hence

$$\sum_{i=1}^{n} P_i \geq 2 \sum_{i=1}^{n} (s_i + t_i)$$

and

$$\sqrt{A_i} \leq \frac{1}{2}(s_i + t_i)$$

Consequently

$$\sum_{i=1}^{n} \sqrt{A_i} \leq \frac{1}{2} \sum_{i=1}^{n} (s_i + t_i) \leq \frac{1}{4} \sum_{i=1}^{n} P_i = 10.$$

Now suppose that $A_i < 0.01$ for $i = 1, 2, \ldots, n$. Then using the above inequality we get

$$1 = \sum_{i=1}^{n} A_i = \sum_{i=1}^{n} \sqrt{A_i}\sqrt{A_i} < \sum_{i=1}^{n} 0.1\sqrt{A_i} \leq 1,$$

a contradiction. Thus $A_i \geq 0.01$ for some $i \in \{1, 2, \ldots, n\}$.

4.15 Broken Lines

3.6.1

(a) It follows from the condition of the problem that the horizontal (vertical) projections of the line segments forming the given broken line do not overlap. Now the solution of the problem follows by the obvious fact that the length of a line segment does not exceed the sum of the lengths of its projections on two perpendicular lines.

(b) Let $ABCD$ be a unit square and O its center. Consider the broken line AEC, where E is a point on the segment OB (Fig. 234).

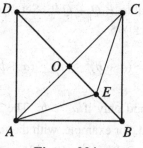

Figure 234.

Its length l takes all values from the interval $[\sqrt{2}, 2)$ as E runs over OB. On the other hand, if E runs over the diagonal AC then the length of the line segment AE takes all values from the interval $(0, \sqrt{2})$.

3.6.2 Assume the contrary. Then some edge $A_1 A_2$ of the broken line P_1 intersects an edge $B_1 B_2$ of the other broken line P_2. The points A_1 and A_2 are not on $B_1 B_2$, because otherwise the distance between two vertices of different broken lines would be less than $\frac{1}{2}$. A similar statement holds for B_1 and B_2, so the quadrilateral $A_1 B_1 A_2 B_2$ is convex. Applying the law of cosines to triangle $A_1 B_1 A_2$ and using the constraints of the problem, we get $\cos \angle A_1 B_1 A_2 \geq 0$, i.e.,

$\angle A_1 B_1 A_2 \leq 90°$. The same is true for the other three angles of $A_1 B_1 A_2 B_2$, and therefore all of them must be 90°. It follows now from the Pythagorean theorem that $(A_1 A_2)^2 = (A_1 B_1)^2 + (B_1 A_2)^2 > 1$, a contradiction.

3.6.3 Suppose the ant begins its path at P_0, stops at $P_1, P_2, \ldots, P_{n-1}$, and ends at P_n (Fig. 235).

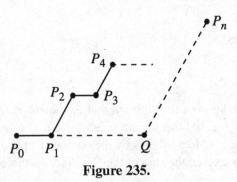

Figure 235.

Note that all segments $P_{2i} P_{2i+1}$ are parallel to each other and that all segments $P_{2i+1} P_{2i+2}$ are parallel to each other. We may then translate all segments so as to form two segments $P_0 Q$ and $Q P_n$ where $\angle P_0 Q P_n = 120°$. Then $P_0 P_n \leq 2r$, and the length of the initial path is equal to $P_0 Q + Q P_n$. Set $P_0 P_n = c$, $P_0 Q = a$, and $Q P_n = b$. Then the law of cosines gives

$$(2r)^2 \geq c^2 = a^2 + b^2 + ab = (a+b)^2 - ab \geq (a+b)^2 - \frac{1}{4}(a+b)^2,$$

so $\frac{4r}{\sqrt{3}} \geq a + b$ with equality if and only if $a = b$. The maximum is therefore $\frac{4r}{\sqrt{3}}$. This maximum can be attained, for example, with the path such that $P_0 P_2$ is a diameter of the circle, and $P_0 P_1 = P_1 P_2 = \frac{2r}{\sqrt{3}}$ (Fig. 236).

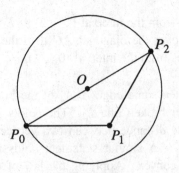

Figure 236.

3.6.4 Let the broken line be formed by n line segments of lengths $l_1, l_2, \ldots, l_n$, respectively. Denote by a_i and b_i the lengths of the orthogonal projections of the ith line segment onto two perpendicular sides of the square. Then $l_i \leq a_i + b_i$, $1 \leq i \leq n$, and therefore

$$1000 = l_1 + \cdots + l_n \leq (a_1 + \cdots + a_n) + (b_1 + \cdots + b_n).$$

We may assume that $a_1 + \cdots + a_n \geq 500$. Then there is a point on the respective side of the square that is covered by the projections of at least 500 line segments of the broken line. Hence the line through that point and perpendicular to the respective side of the square intersects the broken line at least 500 times.

3.6.5 Divide the square into n vertical strips such that each of them contains precisely n of the given n^2 points (the boundary points can be assigned to the left or to the right strip). Then we connect the n points in each strip from up to down and obtain in this way n broken lines. Consider the two broken lines connecting all n^2 points as shown in Fig. 237.

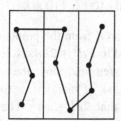 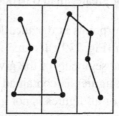

Figure 237.

The union of the line segments connecting the points in consequtive strips is a pair of broken lines whose horizontal projections have lengths less than or equal to 1. Therefore the length of the horizontal projection of one of these broken lines is not greater than

$$1 + (n - 1)(u_1 + u_2 + \cdots + u_n) = n,$$

where u_i is the width of the ith strip. The length of the vertical projection of this broken line is obviously not greater than n and therefore its length is not greater than $2n$.

3.6.6 We shall show that the government has enough money to construct a system of highways connecting all 51 towns. Indeed, we first construct a highway through one of the towns in the vertical direction from the north to the sought boundary of the country. Its length is 1000 km. Then we construct 5 horizontal highways from the west to the east boundary of the country at distances 100 km, 300 km, 500 km, 700 km, and 900 km from its south boundary (Fig. 238).

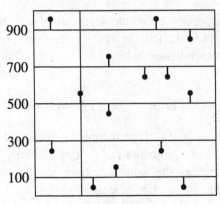

Figure 238.

Then from each of the remaining 50 towns we construct the shortest highway to a horizontal one. The length of any such a highway is not greater than 100 km. The system of highways constructed in this way connects all towns of the country and its total length is not greater than $6 \cdot 1000 + 50 \cdot 100 = 11000$ km.

3.6.7 Consider the set of points at distance less than d from the points of a given line segment of length h (Fig. 239). Its area is equal to $\pi d^2 + 2hd$. Now construct such figures for all n line segments of the given broken line. Since the intersection of any two consecutive figures is contained in a disk of radius d, it follows that the area of the union F of all figures is not greater than $2dl + \pi d^2$. The condition of the problem implies that the set F contains the given unit square and therefore $1 \le 2dl + \pi d^2$, which is equivalent to the inequality $l \ge \frac{1}{2d} - \frac{\pi d}{2}$.

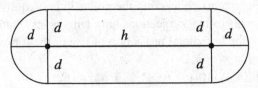

Figure 239.

4.16   Distribution of Points

3.7.1 Let the four vertices of the square be V_1, V_2, V_3, V_4, and let $S = \{P_1, P_2, \ldots, P_n\}$. For a given P_k, we may assume without loss of generality that P_k lies on the side $V_1 V_2$ (Fig. 240).

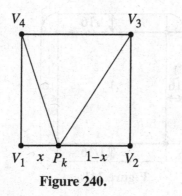

Figure 240.

Writing $x = P_k V_1$, we have

$$\sum_{i=1}^{4} P_k V_i^2 = x^2 + (1-x)^2 + (1+x^2) + (1+(1-x)^2) = 4\left(x - \frac{1}{2}\right)^2 + 3 \geq 3.$$

Hence

$$\sum_{i=1}^{4}\sum_{j=1}^{n} P_j V_i^2 \geq 3n, \quad \text{or} \quad \sum_{i=1}^{4}\left(\frac{1}{n}\sum_{j=1}^{n} P_j V_i^2\right) \geq 3.$$

Thus, if we select the V_i for which $\frac{1}{n}\sum_{j=1}^{n} P_j V_i^2$ is maximized, we are guaranteed it will be at least $\frac{3}{4}$.

3.7.2 Divide the given square into 25 squares of side length $\frac{1}{5}$. Then one of them contains at least 5 of the given 101 points. These 5 points lie in its circumcircle which has radius $\frac{\sqrt{2}}{10} < \frac{1}{7}$.

3.7.3 Suppose that the distance between any two of the given 112 points is at least $\frac{1}{8}$. Consider the disks centered at these points and with radius $\frac{1}{16}$.

Any two of these disks do not intersect and all of them lie in the set A of points shown in Fig. 241. The area of A is equal to $1 + 4 \cdot \frac{1}{16} + \frac{\pi}{16^2}$. Hence

$$1 + \frac{4}{16} + \frac{\pi}{16^2} > \frac{112\pi}{16^2},$$

which is equivalent to $320 > 111\pi$. But this is a contradiction since $111\pi > 333$.

3.7.4 Divide the given unit cube into 8 cubes with edges $\frac{1}{2}$. It is clear that each of them contains exactly one of the given 8 points; otherwise, two of these points are contained in a cube of edge $\frac{1}{2}$, and the distance between them would be less than or equal to $\frac{\sqrt{3}}{2} < 1$, a contradiction.

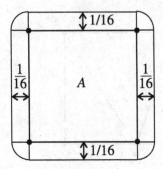

Figure 241.

Now suppose that one of the given points, denote it by M, is not a vertex of the cube. Denote by B the common vertex of the given cube and the cube of edge $\frac{1}{2}$ containing M. Then at least one of the orthogonal projections of M on the three edges through B does not coincide with B; let this edge be AB. Denote by N the point contained in the cube of edge $\frac{1}{2}$ and vertex A and by M_1 and N_1 the orthogonal projections of M and N on AB. Let N_2 be the orthogonal projection of N on the line MM_1. Set $M_1B = d \leq \frac{1}{2}$. Then $M_1N_1^2 + MN_2^2 = MN^2 \geq 1$ and $MN_2 \leq d\sqrt{2}$. Hence $M_1N_1 \geq \sqrt{1-2d^2}$ and we get $d = M_1B \leq AB - M_1N_1 \leq 1 - \sqrt{1-2d^2}$. Therefore $d \leq 1 - \sqrt{1-2d^2}$, which implies that $d \geq \frac{2}{3}$, a contradiction.

3.7.5 Divide the square into 50 horizontal rectangles of height 2. Suppose that one of these rectangles contains at most 7 centers of the given disks. Then the length of the line segment connecting the midpoints of its vertical sides is less than $8 \cdot 10 + 7 \cdot 2 = 94$, a contradiction. Hence each rectangle contains at least 8 centers, and the total number of disks is at least $8 \cdot 50 = 400$.

3.7.6

(a) It is enough to show that the square can be divided into $2(n+1)$ triangles with vertices among the given points $P_1, P_2, \ldots, P_n$ and the vertices of the square. To do this we first divide the square into 4 triangles by connecting P_1 with its vertices. If P_2 lies in the interior of one of these triangles we connect P_2 with its vertices. If P_2 lies on a common side of two triangles we connect P_2 with their opposite vertices. Proceeding in the same way for $P_3, \ldots, P_n$ we finally divide the square into $2(n+1)$ triangles (Fig. 242).

(b) We may assume that no three of the given n points are collinear. Then their convex hull M is a k-gon, $3 \leq k \leq n$. If $k = n$ we divide M into $n-2$ triangles by means of the diagonals through a fixed vertex. If $k < n$ we use the same

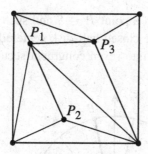

Figure 242.

procedure as in (a) to divide M into $k + 2(n - k - 1) = 2n - k - 2 > n - 2$ triangles. Since the area of M is less than 1 it is clear that in both cases at least one of the obtained triangles has area less than $\frac{1}{n-2}$.

3.7.7 *First solution.* Suppose no such triangle exists. Divide the 25 lattice points $(x, y), x = 0, \pm1, \pm2, y = 0, \pm1, \pm2$, into three rectangular arrays as shown in Fig. 243 (left). If three of the points in $P = \{P_1, P_2, \ldots, P_6\}$ are in the same array, they will determine a triangle with area not greater than 2. Hence each array contains exactly two points in P. By symmetry, each of the arrays in Fig. 243 (right) must contain exactly two points in P. This is a contradiction since P contains only 6 points.

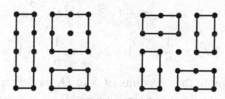

Figure 243.

Second solution. Suppose no such triangle exists. By the pigeonhole principle, at least one row contains two points in P. Then its adjacent rows cannot contain any points in P. Thus the distribution of the points in P among the rows must be $(2, 0, 2, 0, 2)$. By symmetry, this is also their distribution among the columns. Thus we may restrict our attention to the points (x_i, y_i) with $x_i = 0$ or ±2 and $y_i = 0$ or ±2. At least one of $(0, -2)$ and $(0, 2)$ must be in P, and we may assume that $(0, 2)$ is. If $(0, 0)$ is also in P, then these two points determine a triangle of area 2 along with any of the other four points in P. Hence $(0, 0)$ is not in P, and that puts $(0, -2)$, $(-2, 0)$, and $(2, 0)$ in P. However, the inclusion in P of any of the remaining four points will create a triangle of area 2. This is a contradiction.

3.7.8 Take a line l passing through a point $P \in S$ and such that the whole set S lies on one side of l. Let D be the closed half-disk centered at P, of radius $\sqrt{3}$, and diameter on line l. Divide D into seven congruent sectors of angular size $\frac{\pi}{7}$ (Fig. 244).

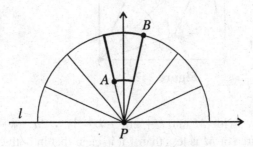

Figure 244.

We are going to show that each sector contains at most one point of S, other than P. To prove this, take a coordinate system such that P is the origin and l the x-axis. Remove from the middle sector of D all points that are within distance 1 from P. What remains, is a "curvilinear quadrilateral," whose most distant points are

$$A\left(-\cos\frac{3}{7}\pi, \sin\frac{3}{7}\pi\right), \quad B\left(\sqrt{3}\cos\frac{3}{7}\pi, \sqrt{3}\sin\frac{3}{7}\pi\right),$$

with

$$AB^2 = (\sqrt{3}+1)^2\cos^2\frac{3}{7}\pi + (\sqrt{3}-1)^2\sin^2\frac{3}{7}\pi$$

$$= 4 - 2\sqrt{3}\cos\frac{1}{7}\pi < 4 - 2\sqrt{3}\cos\frac{1}{6}\pi = 1.$$

It follows that there are no more than 8 points of S in D, including the center P. Delete all these points from S; what remains is a set S' of at least 1972 points. The same procedure can now be performed with respect to S' and we continue this procedure until there are no points left. At each step we kill 8 points of S (having covered them by a half-disk of radius $\sqrt{3}$). Thus the number of steps is not less than $\frac{1980}{8}$, hence not less than 248. The centers of the half-disks constructed in the successive steps constitute a set of at least 248 points, the mutual distance between any two of them exceeding $\sqrt{3}$.

3.7.9 Since no two of the given n points lie on a radial direction (otherwise the distance between them would be less than $\sqrt{2}-1 < 1$), we may order them clockwise. Consider two consecutive points A and B. Set $AO = x$, $BO = y$, $AB = z$, and $\angle AOB = \varphi$, where O is the center of the annulus. Then $1 \le x, y \le \sqrt{2}$, and $z \ge 1$. By the law of cosines we get

$$\cos \varphi = \frac{x^2 + y^2 - z^2}{2xy} \le \frac{x^2 + y^2 - 1}{2xy}.$$

For a fixed x consider the right-hand side of the above inequality as a function of y. This function is increasing for $1 \le y \le \sqrt{2}$ since its first derivative is equal to

$$\frac{y^2 - x^2 + 1}{2xy^2} \ge \frac{2 - x^2}{2xy^2} \ge 0.$$

Hence

$$\frac{x^2 + y^2 - 1}{2xy} \le \frac{x^2 + 1}{2x\sqrt{2}}.$$

On the other hand it is easy to check that

$$\frac{x^2 + 1}{2x\sqrt{2}} \le \frac{3}{4}$$

for $1 \le x \le \sqrt{2}$ and we get that $\cos \varphi \le \frac{3}{4}$. Now the well-known inequality $\cos x \ge 1 - \frac{x^2}{2}$ implies that $\cos \frac{2\pi}{9} \ge 1 - \frac{2\pi^2}{81} > \frac{3}{4} \ge \cos \varphi$, and we conclude that $\varphi > \frac{2\pi}{9}$. This shows that $n < 9$. Thus, the desired largest n is equal to 8, since the 8 vertices of a regular octagon inscribed in the circle of radius $\sqrt{2}$ satisfy the condition of the problem.

3.7.10 The problem can be restated using mathematical terminology as follows:

A set S of ten points in the plane is given, with all the mutual distances distinct. For each point $P \in S$ we mark red the point $Q \in S(Q \ne P)$ nearest to P. Find the least possible number of red points

Note that every red point can be assigned (as the closest neighbor) to at most five points from S. Otherwise, if a point Q were assigned to $P_1, \ldots, P_6$, then one of the angles $P_i Q P_j$ would be not greater than $60°$, in contradiction to $P_i P_j$ being the longest side in the (nonisosceles) triangle $P_i Q P_j$.

Let AB be the shortest segment with endpoints $A, B \in S$. Clearly, A and B are both red. We are going to show that there exists at least one more red point. Assume the contrary, so that for each one of the remaining eight points, its closest neighbor is either A or B. In view of the previous observation, A must be assigned to four points, M_1, M_2, M_3, M_4, and B must be assigned to the remaining four points, N_1, N_2, N_3, N_4. Choose labeling such that the angles $M_i A M_{i+1}$ ($i = 1, , 2, 3$) are successively adjacent, angles $N_i A N_{i+1}$ are so too, the points M_1, N_1 lie on one side of line AB, and M_4, N_4 lie on the opposite side. As before, each angle $M_i A M_{i+1}$ and $N_i A N_{i+1}$ is greater than $60°$. Therefore each one of $\angle M_1 A M_4$ and $\angle N_1 B N_4$ is less than $180°$, and hence $(\angle M_1 AB + \angle N_1 BA) + (\angle M_4 AB + \angle N_4 BA) < 360°$.

At least one of the two sums in the parentheses, say the first one, is less than 180°: $\angle MAB + \angle NBA < 180°$. Here and in the sequel we write M, N instead of M_1, N_1 for brevity.

Since $MA < MB$ and $NB < NA$, the points A, M lie on one side of the perpendicular bisector of AB, and the points B, N lie on the other side. Hence, and because M, N lie on the same side of AB, the points A, B, N, M are consecutive vertices of a quadrilateral. Since AB is the shortest side of the triangle BNA, and since $MA(< MN)$ is not the longest side in the triangle AMN, the angles BNA and ANM are acute. Therefore the internal angle BNM of the quadrilateral $ABNM$ is less than 180°. Similarly, its internal angle NMA is less than 180°. Thus $ABNM$ is a convex quadrilateral (Fig. 245).

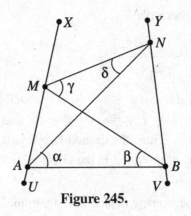

Figure 245.

Choose points U, V, X, Y arbitrarily on the rays MA, NB, AM, BN produced beyond the quadrilateral. The previous condition $\angle MAB + \angle NBA < 180°$ implies the inequalities $\angle UAB + \angle ABV > 180°$ and $\angle XMN + \angle MNY < 180°$.

Define the angles $\alpha = \angle NAB, \beta = \angle ABM, \gamma = \angle BMN, \delta = \angle MNA$. In triangle NAB we have $AB < NB$, so that $\angle ANB < \angle NAB = \alpha$, and thus $\angle ABV = \angle NAB + \angle ANB < 2\alpha$.

In triangle BMN we have $MN > BN$, so that $\angle MBN > \angle BMN = \gamma$, and consequently $\angle MNY = \angle BMN + \angle MBN > 2\gamma$. Analogously, $\angle UAB < 2\beta$ and $\angle XMN > 2\delta$. Hence

$$2\alpha + 2\beta > \angle ABV + \angle UAB > 180° > \angle MNY + \angle XMN > 2\gamma + 2\delta.$$

This yields the desired contradiction because $\alpha + \beta = \gamma + \delta (= \angle AZM$, where Z is the point of intersection of AN and BM).

Thus, indeed, there exists a third red point. The following example shows that a fourth red point need not exist, so that *three* is the minimum sought.

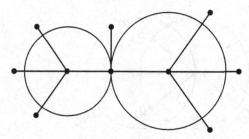

Figure 246.

Example. The two tangent circles in Fig. 246 differ slightly in size.

The acute central angles are greater than 60°. Six points are just a bit outside the circles. The length of the vertical segment is equal to the radius of the bigger circle. Each point has a unique (hence well-defined) closest neighbor, which has to be marked red.

The only three that will be marked red are the two centers and the point of tangency. If some of the (irrelevant) distances happen to be equal, one can slightly perturb the positions of any points without destroying the mentioned properties.

3.7.11 We shall prove the result by induction on n, by means of the following lemma.

Lemma. *If more than two diameters issue from one of the given points, then there is another point from which only one diameter issues.*

Proof. Let A be an endpoint of three (or more) diameters. The other endpoints of these diameters lie on a circle O_A with center A and radius d. Moreover, they all lie on an arc of radian measure $\leq \frac{\pi}{3}$, since otherwise the pair farthest apart will be at a distance $> d$ from each other. Denote the other endpoints of three diameters from A by B_1, B_2, B_3, where B_2 lies between B_1 and B_3 on this arc. With B_2 as a center, draw a circle O_{B_2} with radius d and denote the intersections of O_{B_2} and O_A by P and Q (Fig. 247). We claim that no point of the given set, except A, lies on the circle O_{B_2}. For all points of the major arc PQ (except P and Q) are farther than d away from A, all points on arc PA (including P but not A) are farther than d away from B_1, and all points on arc QA (including Q but not A) are farther than d away from B_2. It follows that B_2A is the only diameter issuing from B_2. Thus, if $k > 2$ diameters issue from A, there is at least one point from which only one diameter issues.

We now proceed by induction on n. For a set of three points, there are obviously at most three diameters. So the assertion of the problem holds for $n = 3$. Suppose it holds for sets of n points with $n = 1, 2, \ldots, m$. We shall show that it then holds for sets of $m + 1$ points.

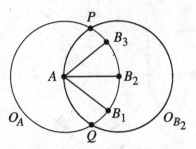

Figure 247.

Consider a set S of $m + 1$ points. We distinguish two cases:

(a) At most two diameters issue from each of the $m+1$ points. Since each diameter has two endpoints, there are at most $\frac{2(m+1)}{2} = m+1$ diameters, so the assertion of the problem holds for S.

(b) There is a point A of S from which more than two diameters issue. Then, by the lemma proved above, there is another point B of S from which only one diameter issues. Now consider the set $S - B$ of m points remaining when B is deleted from S. By the induction hypothesis, $S - B$ has at most m diameters. When B is added to $S - B$, the resulting set S gains exactly one diameter. Hence S has at most $m + 1$ diameters. This completes the proof.

Note that for any $n \geq 3$, there exist sets S of n points in the plane with exactly n diameters. If n is odd, the set S of vertices of a regular n-gon has this property. (See Fig. 248, where $n = 5$.)

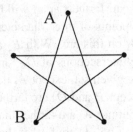

Figure 248.

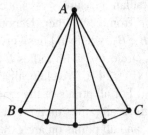

Figure 249.

To get an example that works for all $n \geq 3$, consider Fig. 249. In this figure A, B, C are vertices of an equilateral triangle. The remaining $n - 3$ points are chosen on the circular arc BC with center A.

We note incidentally that Fig. 248 and Fig. 249 illustrate the two cases (a) and (b) occurring in our induction proof.

3.7.12 *First solution.* Consider first the case $n = 5$. We must show that there is at least $\binom{5-3}{2} = 1$ convex quadrilateral. If the convex hull of the five points has four of them on its boundary, they form a convex quadrilateral. If the boundary of the convex hull contains only three of the points, say A, B, C, then the other two, D and E, are inside $\triangle ABC$. Two of the points A, B, C must lie on the same side of the line DE. Suppose for definiteness that A and B lie on the same side of DE, as in Fig. 250. Then $ABDE$ is a convex quadrilateral.

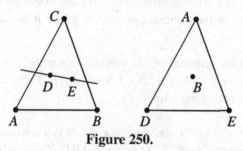

Figure 250.

Consider now the general case $n \geq 5$. With each of the $\binom{n}{5}$ subsets of five of the n points, associate one of the convex quadrilaterals whose existence was demonstrated above. Each quadrilateral is associated with at most $n - 4$ quintuples of points, since there are $n - 4$ possibilities for the fifth point. Therefore there are at least $\frac{\binom{n}{5}}{n-4}$ different convex quadrilaterals in the given set of n points. Now,

$$\frac{1}{n-4}\binom{n}{5} = \frac{n(n-1)(n-2)(n-3)(n-4)}{1 \cdot 2 \cdot 3 \cdot 4 \cdot 5 \cdot (n-4)}$$

$$= \frac{n(n-1)(n-2)}{60(n-4)}\binom{n-3}{2},$$

and it is enough to prove that $n(n-1)(n-2) \geq 60(n-4)$ for $n \geq 5$. This can be seen by forming the difference $n(n-1)(n-2) - 60(n-4) = n^3 - 3n^2 - 58n + 240 = (n-5)(n-6)(n+8)$, and observing that it vanishes for $n = 5$ and $n = 6$ and is positive for all greater n.

Second solution. Choose three points A, B, C of the given set S that lie on the boundary of its convex hull. Then there are $\binom{n-3}{2}$ ways in which two additional points D and E can be selected from S. Once they are chosen, at least two of the points A, B, C must lie on the same side of the line DE. Suppose for definiteness that A and B are on the same side of DE (Fig. 250). Then A, B, D, E are the vertices of a convex quadrilateral. For if not, their convex hull would be a triangle. One of the points A, B would lie inside this triangle, contradicting the fact that A, B, C were chosen to be on the boundary of the convex hull of S. Thus we have found $\binom{n-3}{2}$ convex quadrilaterals whose vertices are among the given points.

4.17 Coverings

3.8.1 Consider the circle of minimum radius R containing the quadrilateral. Then either two vertices of the quadrilateral lie on it and are diametrically opposite, or three vertices lie on it and form an acute triangle. In the first case, $2R \leq 1$, so we certainly have $R < \frac{1}{\sqrt{3}}$. In the second case, let θ be the largest angle of the acute triangle. Then $60° \leq \theta < 90°$ so that $\sin\theta \geq \frac{\sqrt{3}}{2}$. By the extended law of sines, $2R\sin\theta$ is equal to the side of the triangle opposite θ, which is at most 1. Hence $R \leq \frac{1}{\sqrt{3}}$.

3.8.2 Clearly, the unit circles centered at the vertices cover the parallelogram if and only if the unit circles centered at A, B, D cover $\triangle ABD$. To see when this happens, we first prove the following lemma:

Lemma. *Let ABD be an acute triangle, and let r be its circumradius. Then the three circles of radius s centered at A, B, D cover $\triangle ABD$ if and only if $s \geq r$.*

Proof. Since $\triangle ABD$ is acute, its circumcenter O lies inside the triangle. The distances OA, OB, OD are equal to r, so if $s < r$, O does not lie in any of the three circles of radius s centered at A, B, D. It therefore remains only to prove that the circles of radius r centered at A, B, D do indeed cover the triangle. To show this, let L, M, and N be the feet of the perpendiculars from O to the sides BD, DA, AB, respectively (Fig. 251).

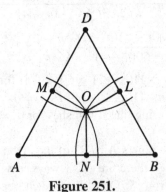

Figure 251.

Then $AN < AO$ and $AM < AO$. Hence the quadrilateral $AMON$ lies inside the circle through O centered at A. Similarly, the quadrilaterals $BLON$ and $DLOM$ lie inside the circles through O centered at B and at D respectively. It follows that $\triangle ABD$ is contained in the union of the three circles. This completes the proof of the lemma.

It is an immediate consequence of the lemma that the unit circles centered at A, B, D cover $\triangle ABD$ if and only if $1 \geq r$. We shall now show that this condition is equivalent to $a \leq \cos \alpha + \sqrt{3} \sin \alpha$.

Let d denote the length of side BD. By the law of cosines,

(1) $$d^2 = 1 + a^2 - 2a \cos \alpha.$$

On the other hand, by the law of sines, $\frac{d}{2r} = \sin \alpha$, and hence $d^2 = 4r^2 \sin^2 \alpha$. Substituting this into (1), we obtain

$$4r^2 \sin^2 \alpha = 1 + a^2 - 2a \cos \alpha.$$

Therefore $r \leq 1$ if and only if

(2) $$4 \sin^2 \alpha \geq 1 + a^2 - 2a \cos \alpha.$$

On the right side of (2), replace the term 1 by $\cos^2 \alpha + \sin^2 \alpha$. Then (2) becomes equivalent to

$$3 \sin^2 \alpha \geq a^2 - 2a \cos \alpha + \cos^2 \alpha = (a - \cos \alpha)^2,$$

and it remains to show that $a - \cos \alpha \geq 0$. To do this, we draw the altitude DQ from D to AB. Since $\triangle ABD$ is acute, Q is inside the segment AB, so $AQ < AB$. But $AQ = \cos \alpha$ and $AB = a$, so $\cos \alpha < a$. This completes the solution.

3.8.3 From the condition, we also know that every point inside or on the triangle lies inside or on one of the six circles.

Define $R = \frac{1}{1+\sqrt{3}}$. Orient triangle ABC so that B is directly to the left of C, and so that A is above BC (Fig. 252).

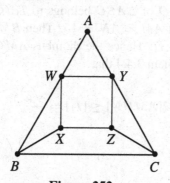

Figure 252.

Draw point W on AB such that $WA = R$, and then draw point X directly below W such that $WX = R$. In triangle WXB, $WB = 1 - R = \sqrt{3}R$ and

$\angle BWX = 30°$, implying that $XB = R$ as well. Similarly draw Y on AC such that $YA = R$, and Z directly below Y such that $YZ = ZC = R$. In triangle AWY, $\angle A = 60°$ and $AW = AY = R$, implying that $WY = R$. This in turn implies that $XZ = R$ and that $WZ = YX = R\sqrt{2}$.

Now if the triangle is covered by six congruent circles of radius r, each of the seven points A, B, C, W, X, Y, Z lies on or inside one of the circles, so some two of them are in the same circle. Any two of these points are at least $R \leq 2r$ apart, so $r \geq \frac{1}{4}(\sqrt{3} - 1)$.

3.8.4 Note first that an equilateral triangle of side length $\frac{3}{2}$ can be covered by means of three equilateral triangles of side length 1. These are the triangles cut from its corners by the lines through its center and parallel to its sides (Fig. 253).

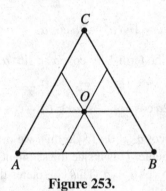

Figure 253.

Now suppose that an equilateral triangle ABC of side length $a > \frac{3}{2}$ is covered by three equilateral triangles T_1, T_2, and T_3 of side lengths 1. Then each of these triangles contains only one of the vertices A, B, C; let $A \in T_1, B \in T_2, C \in T_3$. We may assume that the center O of $\triangle ABC$ belongs to T_1. Consider the points $M \in AB$ and $N \in AC$ such that $AM = AN = \frac{1}{3}a$. Then $BM = CN = \frac{2}{3}a > 1$ and therefore $M \in T_1$ and $N \in T_1$. Hence the rhombus $AMON$ is contained in triangle T_1 and we get from Problem 3.4.4 that

$$\frac{a^2\sqrt{3}}{9} = 2[AMON] \leq [T_1] = \frac{\sqrt{3}}{4}.$$

Thus $a \leq \frac{3}{2}$, a contradiction.

3.8.5

(a) The desired radius R is equal to the circumradius of the equilateral triangle of side length 2, i.e., $R = \frac{2}{\sqrt{3}}$. Indeed, note first that given an equilateral triangle of side length 2 the three unit disks with diameters its sides cover its circumcircle. On the other hand, if three unit disks cover a circle of radius greater than $\frac{2}{\sqrt{3}}$ then one of them contains an arc from this circle of more than $120°$ and hence a chord of length greater than 2, a contradiction.

(b) Assume that $R_1 \leq R_2 \leq R_3$. Using similar arguments as in (a) one can show that if $2R_1, 2R_2, 2R_3$ are side lengths of an acute triangle, i.e., $R_3^2 < R_1^2 + R_2^2$, then its circumradius is the desired one. If $R_3^2 \geq R_1^2 + R_2^2$, the desired radius is equal to R_3.

3.8.6 We shall prove that the desired number is 7.

Note first that a disk D of radius 2 can be covered by 7 unit disks. Indeed, let O be the center of D and let F be a regular hexagon with vertices on its circumference. Then the 6 unit disks with diameters the sides of F together with the unit disk with center O cover D (Fig. 254).

Suppose now that 6 unit disks cover a disk D of radius 2. Since each of them covers no more than $\frac{1}{6}$ part of the circumference of D, it follows that these 6 unit disks form the same configuration as in Fig. 254.

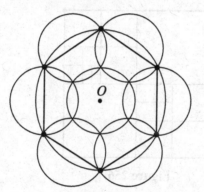

Figure 254.

But then they do not cover the center O of D, a contradiction.

3.8.7 The answer is yes. It is shown in Fig. 255 how one can cover a square of side length $\sqrt{\frac{\sqrt{5}+1}{2}} > \frac{5}{4}$ by means or three unit squares.

3.8.8 We may assume that the side lengths of the given squares are less than 1. Then we cut from each of them the largest square of side length $\frac{1}{2^n}$, where n is a positive integer. Note that given a square of side length $a < 1$, the integer n is uniquely determined by the inequalities $\frac{1}{2^n} \leq a < \frac{1}{2^{n-1}}$. Hence the new squares have side lengths of the form $\frac{1}{2^n}$ and the sum of their areas is at least 1. Now we shall show that one can cover a unit square by means of these new squares. To see

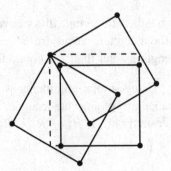

Figure 255.

this we proceed in the following way. We first divide the given square into four squares with side length $\frac{1}{2}$, and put on them all squares from the new collection having side length $\frac{1}{2}$. Suppose that the unit square remains uncovered. Then we divide any of the uncovered squares of side length $\frac{1}{2}$ into 4 squares of side length $\frac{1}{2^2}$ and put on them all squares of side length $\frac{1}{2^2}$ from the new collection.

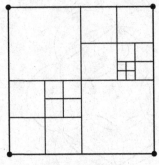

Figure 256.

We may suppose that some of the squares of side length $\frac{1}{2^2}$ remain uncovered and proceed as above until we use all squares from the new collection (Fig. 256). Suppose that after the final step the given square remains uncovered. Since we have used all the squares from the new collection it follows that their total area is less than 1, a contradiction.

Notation

- In a triangle ABC: $a = BC, b = AC, c = AB$; $\alpha = \angle BCA, \beta = \angle ABC$, $\gamma = \angle BCA$;

  r – radius of the incircle; R – radius of the circumcircle;

  O – circumcenter, i.e., the center of the circumcircle of the triangle;

  H – orthocenter, i.e., the intersection point of the altitudes in the triangle;

  G – centroid, i.e., the intersection point of the medians of the triangle;

  m_a, m_b, m_c – the lengths of the medians through A, B, and C, respectively;

  h_a, h_b, h_c – the lengths of the altitudes through A, B, and C, respectively;

- $s = \frac{a+b+c}{2}$ – semiperimeter of $\triangle ABC$

- $[ABC...]$ – the area of the polygon $ABC...$

- $\text{Vol}(P)$ – volume of the polyhedron P

- $\overrightarrow{AB}$ – the vector determined by the points A and B

- $\overrightarrow{AB} \cdot \overrightarrow{CD} = (\overrightarrow{AB}, \overrightarrow{CD}) = AB \cdot CD \cdot \cos \alpha$ – dot (inner) product of the vectors $\overrightarrow{AB}$ and $\overrightarrow{CD}$. Here α is the angle between the two vectors.

Glossary of Terms

- *Circle of Apollonius*: The locus of a point that moves so that the ratio of its distances from two given points is constant is a circle (or a line).

- *Arithmetic mean–geometric mean inequality*:

$$\frac{x_1 + x_2 + \cdots + x_n}{n} \geq \sqrt[n]{x_1 x_2 \cdots x_n}$$

  for any nonnegative real numbers $x_1, \ldots, x_n$. Equality holds if and only if $x_1 = x_2 = \cdots = x_n$.

- *Cauchy–Schwarz inequality*: For any real numbers $x_1, x_2, \ldots, x_n$ and $y_1, y_2, \ldots, y_n$,

$$(x_1^2 + x_2^2 + \cdots + x_n^2)(y_1^2 + y_2^2 + \cdots + y_n^2) \geq (x_1 y_1 + x_2 y_2 + \cdots + x_n y_n)^2,$$

  with equality if and only if x_i and y_i are proportional, $i = 1, 2, \ldots, n$.

- *Centroid of a triangle*: The intersection point of its medians.

  More generally, if $A_1, A_2, \ldots, A_n$ are points in the plane or in space, their *centroid G* is the unique point for which

$$\overrightarrow{GA_1} + \overrightarrow{GA_2} + \cdots + \overrightarrow{GA_n} = \vec{0}.$$

- *Centroid of a tetrahedron*: The intersection point of its medians, i.e. the segments connecting its vertices with the centroids of the opposite faces. (See also the above.)

- *Ceva's theorem*: If AD, BE, and CF are concurrent cevians (a cevian is a segment joining a vertex of a triangle with a point on the opposite side) of a triangle ABC, then (i) $BD \cdot CE \cdot AF = DC \cdot EA \cdot FB$. Conversely, if AD,

BE, and CF are three cevians of a triangle ABC such that (i) holds, then the three cevians are concurrent.

- *Circumcenter*: Center of the cir cumscribed circle or sphere.

- *Circumcircle*: Circumscribed circle.

- *Chebyshev's inequality*: For any real numbers $x_1 \leq x_2 \leq \cdots \leq x_n$ and $y_1 \leq y_2 \leq \cdots \leq y_n$,

$$\left(\frac{1}{n} \sum_{i=1}^{n} x_i \right) \left(\frac{1}{n} \sum_{i=1}^{n} y_i \right) \leq \frac{1}{n} \sum_{i=1}^{n} x_i \, y_i,$$

with equality if and only if $x_1 = x_2 = \cdots = x_n$ or $y_1 = y_2 = \cdots = y_n$.

- *Convex function*: A function $f(x)$ defined on an interval I is said to be convex if

$$f\left(\frac{x+y}{2} \right) \leq \frac{f(x) + f(y)}{2}$$

for any $x, y \in I$. If the second derivative $f''(x)$ exists and $f''(x) \geq 0$ for all $x \in I$, then f is convex on I.

- *Convex hull* of a set F (in the plane or in space): The smallest convex set containing F.

- *Convex polygon*: A polygon in the plane that lies on one side of each line contaning a side of the polygon.

- *Convex polyhedron*: A polyhedron in space that lies on one side of each plane contaning a face of the polyhedron.

- *Cyclic polygon*: A polygon that can be inscribed in a circle.

- *Dilation (homothety)* with center O and coefficient $k \neq 0$ (in the plane or in space): A transformation that assigns to every point A the point A' such that $\overrightarrow{OA'} = k \cdot \overrightarrow{OA}$.

- *Euler's formula*: If O and I are the circumcenter and the incenter of a triangle with inradius r and circumradius R, then $OI^2 = R^2 - 2Rr$.

- *Euler's line*: The line through the centroid G, the orthocenter H, and the circumcenter O.

- *Incenter*: Center of inscribed circle or sphere.

- *Incircle*: Inscribed circle.

- *Jensen's inequality*: If $f(x)$ is a convex function on an interval I, then

$$f\left(\frac{a_1 + a_2 + \cdots + a_n}{n}\right) \leq \frac{f(a_1) + f(a_2) + \cdots + f(a_n)}{n}$$

for any positive integer n and for any choice of $a_1, \ldots, a_n \in I$.

- *Heron's formula*: The area F of an arbitrary triangle with sides a, b, and c and *semiperimeter* $s = \frac{a+b+c}{2}$ is

$$F = \sqrt{s(s-a)(s-b)(s-c)}.$$

- *Law of Sines*:

$$\frac{BC}{\sin \alpha} = \frac{CA}{\sin \beta} = \frac{AB}{\sin \gamma} = 2R$$

in any triangle ABC with circumradius R and angles α, β, and γ, respectively.

- *Law of cosines*:

$$BC^2 = AC^2 + BC^2 - 2AC \cdot BC \cdot \cos \alpha$$

in any triangle ABC.

- *Leibniz's formula*: Let G be the centroid of a set of points $\{A_1, A_2, \ldots, A_n\}$ in the plane (space).

Then for any point M in the plane (space) we have

$$MA_1^2 + MA_2^2 + \cdots + MA_n^2 = n \cdot MG^2 + GA_1^2 + GA_2^2 + \cdots + GA_n^2.$$

- *Median formula*:

$$m_c^2 = \frac{1}{4}(2a^2 + 2b^2 - c^2).$$

- *Minkowski's inequality*: For any real numbers $x_1, x_2, \ldots, x_n, \; y_1, y_2, \ldots,$ $y_n, \ldots, z_1, z_2, \ldots, z_n,$

$$\sqrt{x_1^2 + y_1^2 + \cdots + z_1^2} + \sqrt{x_2^2 + y_2^2 + \cdots + z_2^2} + \cdots + \sqrt{x_n^2 + y_n^2 + \cdots + z_n^2} \geq$$
$$\sqrt{(x_1 + x_2 + \cdots + x_n)^2 + (y_1 + y_2 + \cdots + y_n)^2 + \cdots + (z_1 + z_2 + \cdots + z_n)^2},$$

  with equality if and only if $x_i, y_i, \ldots, z_i$ are proportional, $i = 1, 2, \ldots, n$.

- *Orthocenter of a triangle*: The intersection point of its altitudes.

- *Pick's theorem*: Given a non-self-intersecting polygon P in the coordinate plane whose vertices are at lattice points, let B denote the number of lattice points on its boundary and let I denote the number of lattice points in its interior. Then the area of P is given by the formula $I + B/2 - 1$.

- *Pigeonhole principle*: If n objects are distributed among k boxes and $k < n$, then some box contains at least two objects.

- *Power-of-a-point theorem*:

  (a) If AB and CD are two chords in a circle that intersect at a point P (which may be inside, on, or outside the circle), then $PA \cdot PB = PC \cdot PD$.

  (b) If the point P is outside a circle through points A, B, and T, where PT is tangent to the circle and PAB a secant, then $PT^2 = PA \cdot PB$.

- *Ptolemy's theorem*: If a quadrilateral $ABCD$ is cyclic, then $AB \cdot CD + BC \cdot AD = AC \cdot BD$.

- *Regular polygon*: A convex polygon all of whose angles are equal and all of whose sides have equal lengths.

- *Regular tetrahedron*: A tetrahedron all edges of which have equal lengths.

- *Rhombus*: A parallelogram with sides of equal length.

- *Root mean square–arithmetic mean inequality*:

$$\left(\frac{x_1 + x_2 + \cdots + x_n}{n} \right)^2 \leq \frac{x_1^2 + x_2^2 + \cdots + x_n^2}{n},$$

for any real numbers $x_1, \ldots, x_n$, where equality holds if and only if $x_1 = x_2 = \cdots = x_n$.

- *Rotation* through an angle α (counterclockwise) about a point O in the plane is the transformation of the plane that assigns to any point A the point A' such that $OA = OA'$, $\angle AOA' = \alpha$, and the triangle OAA' is counterclockwise oriented.

- *Simson's theorem:* For any point P on the circumcircle of a triangle ABC, the feet of the perpendiculars from P to the sides of ABC all lie on a line called the *Simson line* of P with respect to triangle ABC.

- *Trigonometric identities:*

$$\sin^2 \alpha + \cos^2 \alpha = 1,$$

$$\tan \alpha = \frac{\sin \alpha}{\cos \alpha},$$

$$\cot \alpha = \frac{\cos \alpha}{\sin \alpha},$$

$$\csc(\alpha) = \frac{1}{\sin \alpha};$$

addition and subtraction formulas:

$$\sin(\alpha \pm \beta) = \sin \alpha \, \cos \beta \pm \cos \alpha \, \sin \beta,$$

$$\cos(\alpha \pm \beta) = \cos \alpha \, \cos \beta \mp \sin \alpha \, \sin \beta,$$

$$\tan(\alpha \pm \beta) = \frac{\tan \alpha \pm \tan \beta}{1 \mp \tan \alpha \, \tan \beta};$$

double-angle formulas:

$$\sin(2\alpha) = 2 \sin \alpha \, \cos \alpha,$$

$$\cos(2\alpha) = 2 \cos^2 \alpha - 1 = 1 - 2 \sin^2 \alpha,$$

$$\tan(2\alpha) = \frac{2 \tan \alpha}{1 - \tan^2 \alpha};$$

triple-angle formulas:

$$\sin(3\alpha) = 3 \sin \alpha - 4 \sin^3 \alpha,$$

$$\cos(3\alpha) = 4 \cos^3 \alpha - 3 \cos \alpha,$$

$$\tan(3\alpha) = \frac{3\tan\alpha - \tan^3\alpha}{1 - 3\tan^2\alpha};$$

half-angle formulas:

$$\sin\alpha = \frac{2\tan\frac{\alpha}{2}}{1 + \tan^2\frac{\alpha}{2}},$$

$$\cos\alpha = \frac{1 - \tan^2\frac{\alpha}{2}}{1 + \tan^2\frac{\alpha}{2}},$$

$$\tan\alpha = \frac{2\tan\frac{\alpha}{2}}{1 - \tan^2\frac{\alpha}{2}};$$

sum-to-product formulas:

$$\sin\alpha + \sin\beta = 2\sin\frac{\alpha+\beta}{2}\cos\frac{\alpha-\beta}{2},$$

$$\cos\alpha + \cos\beta = 2\cos\frac{\alpha+\beta}{2}\cos\frac{\alpha-\beta}{2},$$

$$\tan\alpha + \tan\beta = \frac{\sin(\alpha+\beta)}{\cos\alpha\,\cos\beta};$$

difference-to-product formulas:

$$\sin\alpha - \sin\beta = 2\sin\frac{\alpha-\beta}{2}\cos\frac{\alpha+\beta}{2},$$

$$\cos\alpha - \cos\beta = -2\sin\frac{\alpha-\beta}{2}\sin\frac{\alpha+\beta}{2},$$

$$\tan\alpha - \tan\beta = \frac{\sin(\alpha-\beta)}{\cos\alpha\,\cos\beta};$$

product-to-sum formulas:

$$2\sin\alpha\,\cos\beta = \sin(\alpha+\beta) + \sin(\alpha-\beta),$$

$$2\cos\alpha\,\cos\beta = \cos(\alpha+\beta) + \cos(\alpha-\beta),$$

$$2\sin\alpha\,\sin\beta = -\cos(\alpha+\beta) + \cos(\alpha-\beta).$$

Bibliography

[1] T. Andreescu and Z. Feng, *103 Trigonometry Problems: From the Training of the USA IMO Team*, Birkhäuser, Cambridge, 2005.

[2] T. Andreescu and D. Andrica, *Complex Numbers from A to ... Z*, Birkhäuser, Cambridge, 2005.

[3] V. Boltyanskii and I. Gohberg, *Decomposition of Figures into Smaller Parts*, The University of Chicago Press, Chicago, 1980.

[4] W. Blaschke, *Kreis and Kugel*, Walter de Gruyter & Co., Berlin, 1956.

[5] W. J. Blundon, *Inequalities associated with the triangle*, Canadian Mathematical Bulletin, B(1965), pp. 615–626.

[6] R. Courant and H. Robbins, *What Is Mathematics?* Oxford University Press, London, 1941.

[7] M. Goldberg, *On the original Malfatti problem*, Math. Magazine, **5, 40** (1967), pp. 241–247.

[8] R. A. Johnson, *Advanced Euclidean Geometry*, Dover, 1960.

[9] G. H. Hardy, J. E. Littlewood, G. Pólya, *Inequalities*, Cambridge University Press, Cambridge, 1934.

[10] N.D. Kazarinoff, *Geometric Inequalities*, Random House, New York, 1961.

[11] G. H. Lawden, *Two related triangle maximization problems*, Math. Gazette **66** (1982), pp. 116–120.

[12] H. Lob and H. W. Richmond, *On the solutions of the Malfatti's problem for a triangle*, Proc. London Math. Soc., **2, 30**(1930), pp. 287–304.

[13] G. Malfatti, *Memoria sopra un problema sterotomico*, Memorie di matematica e fisica della Società Italiana delle Scienze, **10, 1**(1803), pp. 235–244.

[14] D. S. Mitrinovic, *Analytic Inequalites*, Springer-Verlag, Heidelberg, 1970.

[15] D. S. Mitrinović, J. E. Pečarić, V. Volenec, *Recent Advances in Geometric Inequalities*, Kluwer Academic Publishers, Dordrecht, 1989.

[16] O. Mushkarov and L. Stoyanov, *Extremal Problems in Geometry* (in Bulgarian), Narodna Prosveta, Sofia, 1989.

[17] P.S. Modenov and A.S. Parkhomenko, *Geometric Transformations*, Academic Press, New York, 1965.

[18] G. Polya, *Mathematics and Plausible Reasoning, Vol. I Induction and analogy in mathematics*, Princeton University Press, Princeton, New Jersey, 1954.

[19] V. M. Tihomirov, *Stories about Maxima and Minima*, AMS, Providence, RI, 1990.

[20] L. F. Tóth, *Lagerungen in der Ebene auf der Kugel und im Raum*, Die Grundlehren der Mathematischen Wissenschaften in Einzeldarstellungen, Band 65, Springer-Verlag, Berlin 1972.

[21] A. S. Winsor, *Modern Higher Plane Geometry*, Christopher Publishing House, Boston, 1941.

[22] I. M. Yaglom, *Geometric Transformations*, Random House, New York, 1962.

[23] V. A. Zalgaller and G. A. Loss, *A solution of the Malfatti problem* (in Russian), Ukrainian Geometric Sbornik, **34** (1991), pp. 14–33.